# Foreword

This report demonstrates the Agency's determination not only to provide information to support better policy-making, but also to gather and disseminate 'best-practice' information that can help actors on the ground find better solutions for the environment and sustainable development.

Renewable energy is increasingly becoming a vital component of the energy mix of the 21st century, and its successful penetration of the market is already underway in many EU Member States.

This publication examines examples of successful implementation of renewable energy projects through case studies and analysis of EU Member State policies and activities. The experiences described may help policy implementers in the Member States to learn from each other's experiences. We hope these success stories will breed further successes.

I believe that this publication could be useful not only to EU decision makers and policy implementers but also to all countries interested in increasing their levels of renewable energy sources.

The report sheds light on the factors that led to successful implementation of some renewable energy technologies in some EU Member States and emphasizes the wide range of factors that can determine whether projects succeed or not. All examples studied are reported following the same methodology and format, thus creating a framework for reporting on success stories that others could also use as a communication tool for promoting renewables.

This document is also available in electronic format on the Agency's main web site at http://reports.eea.eu.int/ as well as in EnviroWindows, an EEA web site for gathering and disseminating information on environmental management and best practices, aimed at businesses, local authorities and their stakeholders (http://ewindows.eu.org/LocalAuthorities/renewables.pdf). It is my hope that this report will become the seed for the creation of a clearing house for experiences in how best to promote renewable energies at many levels, from national to local. The Agency is offering its electronic space and expertise to anybody who would like to pick up the gauntlet and create a permanent site where success stories on implementation of renewable energies are reported in a standardised way and updated when new data become available.

I would like to thank the manager of this report, Aphrodite Mourelatou, for the valuable work she put into this project; Ecotec Research and Consulting Ltd, who assisted with writing the report; Eurostat, which provided the necessary data; and all others who contributed to the report or reviewed it.

Domingo Jiménez-Beltrán
Executive Director

# Contents

Environmental issue report | No 27

# Renewable energies: success stories

Prepared by:
Ecotec Research and Consulting Ltd.
and
Aphrodite Mourelatou
European Environment Agency

European Environment Agency

Cover design: Brandenborg a/s
Layout: Brandenborg a/s
Photos: Dr.-Ing.habil. Volker Quaschning, Solar System Analysis

**Legal notice:**
The contents of this publication do not necessarily reflect the official opinions of the European Commission or other European Communities institutions. Neither the European Environment Agency nor any person or company acting on the behalf of the Agency is responsible for the use that may be made of the information contained in this report.

A great deal of additional information on the European Union is available on the Internet.
It can be accessed through the Europa server (http://europa.eu.int).

Luxembourg: Office for Official Publications of the European Communities, 2001

ISBN 92-9167-407-9

© EEA, Copenhagen, 2001

*Printed in Germany*

Printed on recycled and chlorine-free bleached paper

European Environment Agency
Kongens Nytorv 6
DK-1050 Copenhagen K
Denmark
Tel: (45) 33 36 71 00
Fax: (45) 33 36 71 99
E-mail: eea@eea.eu.int
Internet: http://www.eea.eu.int

# Executive summary

**Background**

This study presents a series of examples of successful renewable energy development in EU Member States. By analysing these cases it seeks to understand the factors that influence the success of renewable energy schemes and to facilitate their widespread penetration in the Member States. The study aims to provide policy-makers and policy implementers with an opportunity to learn from each others' experiences, and thereby to contribute to EU and Member State efforts to meet their renewable energy indicative targets for 2010.

**Identifying progress in renewable energy deployment**

The study evaluates the rate of penetration in each EU Member State of four renewable energy sources — photovoltaics, solar thermal, wind and biomass (as biomass power, biomass district heating and biofuels) — between 1993 and 1999. For consistency purposes only data from Eurostat (the statistical office of the European Communities) have been used.

For the purposes of this study, two selection criteria have been applied to identify those Member State/technology combinations where renewable energy has penetrated to a greater degree of success than in others:

- An absolute increase equivalent to at least 10 % of the total EU-wide increase in renewable energy output of that particular technology over the period 1993–99. The 10 % threshold has been selected to identify those Member States that made the greatest contribution to the increase in renewable energy output of each technology in the EU.

- A percentage increase of renewable energy output of the examined technology greater than the EU percentage increase for that technology between 1993–1999. This compares the percentage increase of each renewable energy source with the EU-wide percentage increase, and identifies those Member State/technology combinations which exceeded the EU-wide figure.

This approach enables the identification of those Member State/technology combinations where exploitation of the technology was well established in 1993 and continued to expand over the next six years. It also highlights Member States where a technology was in an initial stage of development at the start of the six-year period and achieved a rapid rate of increase in penetration, but still only provides a limited quantity of energy output.

It should be noted that the criteria cannot be used to directly benchmark a country's performance in promoting renewable energies. The study focuses on positive changes during the period studied, and does not compare absolute levels of penetration.

The results of applying these two criteria are summarised in the table below.

**Trends in renewable energy penetration, 1993–99**

Source: Eurostat.

| Technology: Selection criteria (see note 1): | Photo-voltaics | | Solar thermal | | Wind | | Biomass: power | | Biomass: district heating (1993–98) | | Biomass: biofuels (see note 2) |
|---|---|---|---|---|---|---|---|---|---|---|---|
| Austria | | | ✓ | ✓ | | | | ✓ | ✓ | ✓ | ✓ |
| Belgium | | | | | | | | ✓ | | | |
| Denmark | | | | ✓ | ✓ | | | ✓ | | | |
| Finland | | | | ✓ | | ✓ | ✓ | | | | |
| France | | | | | | ✓ | | | | ✓ | ✓ |
| Germany | ✓ | ✓ | ✓ | ✓ | ✓ | ✓ | | ✓ | | | ✓ |
| Greece | | | ✓ | | | | | | | | |
| Ireland | | | | ✓ | | ✓ | | | | | |
| Italy | | | | ✓ | | ✓ | | ✓ | | ✓ | ✓ |
| Luxembourg | | | | | | | | | | | |
| Netherlands | | ✓ | | ✓ | | | | | | | |
| Portugal | | | | | | ✓ | | | | | |
| Spain | ✓ | ✓ | | | ✓ | | | ✓ | | | |
| Sweden | | | | | | ✓ | ✓ | | ✓ | ✓ | |
| UK | | | | ✓ | | | | | | | |

Biomass district heating refers to heat output from heat plants only.

Note 1: Two criteria for selection are used:
    ✓ (left) represents a contribution of at least 10 % of the total EU increase in absolute terms, 1993–99;
    ✓ (right) represents a percentage increase greater than the EU percentage increase, 1993–99.
Note 2: Biofuels only:
    ✓ represents those Member States which indicate that they use biofuels (most do not).

The results show wide variations between the 15 Member States (see Table 3 on page 21).

For most technologies, only two or three Member States contributed more than 10 % (each) of the total new output from the technology:

— Germany and Spain contributed 78 % of new EU output from photovoltaics;
— Austria, Germany and Greece contributed 80 % of new solar thermal installations;
— Denmark, Germany and Spain contributed 80 % of new wind output;
— Finland and Sweden contributed 60 % of new generation from biomass-fuelled power stations (including biomass combined heat and power stations);
— Austria and Sweden dominated the increase in output from biomass district heating installations;
— only four Member States use biofuels to any significant extent; France is the market leader, producing about 40 % of the total.

More Member States, however, were starting to expand their rates of exploitation of certain technologies, and had a percentage increase greater than the EU figure. Indeed, some Member States with low initial levels of renewable energy use demonstrated rapid growth rates even though the actual increase in output achieved was small over the period.

Only a small number of Member State/technology combinations met both criteria. Germany achieved the greatest levels of new renewable energy penetration over the period and met both criteria for all of the technologies studied except biomass. There were also positive combinations in Austria, Spain and Sweden.

### Success factors
The study drew on the considerable amount of previous work carried out at EU and Member State level on the barriers that hinder the implementation of renewable energies. It identifies a number of potential success factors likely to have a positive influence on the development of renewable energy technologies.

The study examines the influence of these potential success factors in a series of case studies selected from the Member State/technology combinations that have been identified by applying the selection criteria summarised in the table above. In particular, it examines all Member State/technology combinations that met both of the criteria for successful penetration and the most interesting Member State/technology combinations meeting one of the criteria. Where possible and appropriate, representative examples of renewable energy projects have been used in the case studies.

The following conclusions summarise the results of the analysis carried out, based on the case studies.

No single factor was identified as being of overwhelming significance. It is rather the cumulative benefits of a series of supportive measures that determine the extent to which a renewable technology is successfully exploited. There are, however, certain essential components of successful renewable energy implementation that help to create an environment in which renewable energy exploitation can succeed.

These are described below under the following seven sub-headings: political; legislative; fiscal; financial; administrative; technological development; information, education and training:

### Political support

- The EU countries which showed a rapid expansion of renewable energy during the 1990s are most commonly those with long-established policies in support of renewable energy in general or of a particular renewable energy.

Regional energy policies can also contribute towards encouraging renewable energy development. For Member States with a high degree of regional autonomy, such as Austria, Germany and Spain, many regional authorities have brought forward energy plans that are even more supportive towards renewable energy than those at the national level.

### Legislative support

- For electricity from renewable sources, the **feed-in law system** has given a great impetus to renewable energy developments, in particular wind energy. This system combines commercially favourable guaranteed feed-in tariffs with an obligation on utilities to purchase renewable electricity at these tariffs.

Denmark, Germany and Spain (all countries using the feed-in law system) contributed 80 % of new wind energy output in the EU-15 over the period 1993–99.

Biomass use in power stations also benefited from feed-in laws, particularly in these same three countries. However, the expansion of biomass power has not been as rapid as that of wind. Biomass use also increased significantly in some Member States without the support of a feed-in mechanism (Finland and Sweden). Successful biomass development benefits from feed-in tariffs but is also closely linked with other success factors, especially the availability of financial support (see below).

The use of photovoltaics expanded significantly in those Member States that provided a high level of support to projects through feed-in arrangements — the most successful being Germany and Spain. However, successful penetration occurred where the feed-in support initiatives were implemented together with capital subsidy programmes to encourage uptake of the technology (again Germany and, to a lesser extent, Spain).

The main alternative to the feed-in mechanism is the **competitive tendering process**. This system was chosen by Ireland and the UK ([1]) to support a range of technologies, including wind and biomass, and by France ([2]) to support wind energy. Developers are guaranteed that, if they win the tender, their power will be purchased at the price they bid in their proposal. This resulted in an increase in capacity of a range of renewable energy technologies in these

---

(1)  The UK is replacing the competitive tendering system with a Renewables Obligation, under which suppliers are legally obliged to provide an increasing proportion of their supplies from renewable sources.
(2)  From June 2001 France replaced competitive tendering with a system similar to the feed-in law.

Member States, but not to the same extent as was achieved where feed-in arrangements were available. Compared with feed-in arrangements in other Member States, the competitive nature of the tendering system has offered fewer guarantees to developers that they will receive an acceptable tariff rate for their project.

- Renewable energy power generators need grid access to be able to distribute the electricity generated. This requires the establishment of transparent and reasonable charging structures so they can operate successfully within the electricity supply system.

Member States that took the biggest steps to address problems of grid access achieved the greatest levels of renewable electricity penetration during the 1990s, especially for smaller-scale renewable energy projects.

Grid access is an important component of the recently adopted EU directive ([3]) on renewable energy in the internal electricity market. The directive requires Member States to take the necessary measures to guarantee the transmission and distribution of electricity produced from renewable sources and encourages such electricity to be given priority access to the grid.

### Fiscal support
- Fiscal (taxation) measures are increasingly being used as a mechanism to reward the environmental benefits of renewable energy compared with energy generated from fossil sources.

During the period of this study, at least six Member States — Austria, Denmark, Finland, Italy, Netherlands and Sweden — put in place some form of energy-related taxation which penalises the use of fossil fuels or other environmentally damaging activities. In Sweden, the introduction of taxes on carbon dioxide emissions and energy from which biomass is exempted helped the expansion of both biomass district heating and biomass combined heat and power plants. The taxes made other options, in particular coal-fired district heating and coal-fired combined heat and power plants, more expensive.

An alternative fiscal approach developed in other Member States is to allow various tax exemptions or reductions for individuals or companies who invest in or use renewable-related products or services. The installation of solar thermal water-heating systems has been stimulated by this approach (in Greece, for example). Individual investors in wind energy benefit from tax exemptions in Germany and Sweden, while Dutch companies benefit from accelerated depreciation for investment in energy-saving schemes that include renewable projects.

Biofuels for transport represent a special example of the role of fiscal initiatives to support renewable energy. Some Member States took advantage of opportunities to apply lower fuel excise duty rates to support biofuels. France in particular used this option to stimulate the growth of the EU's largest biofuels industry.

### Financial support
- The capital costs of renewable energy projects, which are often high, can be a significant barrier to development, especially for newer technologies. Subsidies or favourable loans for renewable energy developments are common where successful penetration occurs.

**Wind** energy has become increasingly cost-competitive over the 1990s, and grants are required less often where feed-in arrangements are available. Where such arrangements are not available, subsidies are still the main mechanism for successfully supporting wind energy schemes.

**Biomass** installations usually still need capital subsidy to be financially viable as they have high capital costs. This is true even in countries with feed-in arrangements. For example, capital

---

(3)   Directive of the European Parliament and of the Council on the promotion of electricity from renewable energy sources in the internal electricity market (2001/77/EC).

grants are provided for biomass power schemes in Germany to augment feed-in tariffs, but for wind power only limited grants are currently provided.

**Photovoltaics (PV)** is still an immature technology and projects need considerable subsidy. Member States showing significant increases in PV use are those that have established subsidy support mechanisms, usually in association with feed-in tariff arrangements.

**Solar thermal installations** are not always able to compete on costs with fossil fuel heating sources and require subsidies to achieve a high level of uptake. Substantial subsidies were made available to households or industry in the three Member States showing the greatest level of development of solar thermal power over the period of this study (Austria, Germany and Greece).

### Administrative support

• Successful replication of renewable energy projects can be achieved on a wide scale only where there is active support for renewable energy at the level at which individual projects are brought forward for approval. In most cases this is the local or regional level. Administrative support at national level is also an important component for success.

The case studies identified a wide range of ways in which local or regional administrations or municipalities have successfully supported renewable energy uptake, including:

— developing local or regional targets for renewable energy uptake;
— providing planning guidance for locating renewable projects;
— providing funding support for local renewable energy projects;
— improving building regulations to stimulate uptake of photovoltaics or solar thermal installations;
— implementing district heating schemes;
— supporting the development of new indigenous manufacturing capacity.

Some of the projects studied received a strong level of public sector support at regional or local level. This shows the regional or local authorities' appreciation of the benefits that the projects would bring to the local population not only in terms of the environment but also in terms of economic development and employment.

### Technological development

• The development of renewable energy technologies requires support at all stages — research, demonstration and implementation — to help achieve strong and competitive indigenous industry capabilities in renewable energy.

Public sector funding to develop renewable energy has been provided over the past 20 years, both by the European Commission and at Member State level. This support has helped to achieve considerable cost reductions and technological improvements. Member States that have focused national research funding support towards specific technologies are now reaping the rewards of this investment. For example, Denmark now has the world's leading wind industry, Finland and Sweden have strong capabilities in biomass technologies, and Germany and the Netherlands are home to photovoltaic cell manufacturing.

### Information, education and training

• Activities that raise awareness of the benefits of renewable energy among the general public are a vital component of national, regional and local renewable energy support programmes.

The success of a renewable energy project, and its subsequent replication, ultimately depends on its public acceptability at local level. At this level, communicating the non-energy benefits is an important component in the acceptability of renewable energy, especially its role in providing revenue and local jobs. Cooperative participation in a project can be a successful way of involving the local community in a new renewable energy development. The role of the developer could also be important for public acceptance and requires that developers work with the local community to provide information about the nature of new renewable energy projects and their potential benefits.

**Energy agencies** at local or regional level are one of the most successful initiatives to help raise public awareness of the benefits of renewable energy and increase public acceptance of new renewable energy developments. Their role is to stimulate the expansion of renewable energy and energy efficiency in their area through public and private sector initiatives and local community involvement.

# 1. Introduction

**Photo:** Volker Quaschning

EU targets for renewable energy will be achieved through actions at the national, regional and, increasingly, local levels. Member States differ considerably in terms of the contribution made by renewable energy to their energy consumption. Member States also have very variable levels of exploitation of different types of renewable energy sources.

This study provides policy-makers and policy implementers with background information and analysis into the successful penetration of a number of renewable energy technologies in EU Member States. The report examines examples of successful penetration, through case studies and analysis of Member State policies and activities. It attempts to shed light on the factors which led to successful implementation of renewable energy in some Member States and in some technologies. The study aims, through the provision of this information, to help policy implementers learn from each others' experiences and contribute to the efforts to meet indicative renewable energy targets.

**Section 2** provides background information on what renewable energy is, the importance of renewable energy, renewable energy targets and the rationale behind the choice of the renewable energies covered in this study.

The renewable energy sources covered are:

— solar photovoltaics
— solar thermal
— wind
— biomass energy (as biomass power, biomass district heating and biofuels).

**Section 3** develops and applies, for the purposes of this study, a set of selection criteria in order to identify those Member State/technology combinations where renewable energy has penetrated to a greater degree of success than in others.

**Section 4** presents the wide range of factors that can influence the likely successful penetration of renewable energy technologies in different Member States, drawing on the considerable amount of previous work carried out at EU and Member State levels on the barriers that hinder implementation of renewable energies.

**Section 5 and Annex 1** present examples of successful Member State/technology combinations in the light of the potential success factors identified previously.

**Section 6** draws together some of the key issues and potential success factors identified in the examples of successful Member States/technology combinations.

**Section 7** draws some conclusions on factors which may contribute to successful penetration of renewables in the Member States.

# 2. Background

**Photo:** Volker Quaschning

## 2.1. The importance of renewable energy

Renewable energy is the term used to describe a wide range of naturally occurring, replenishable energy sources — in particular, sun, wind, water and a range of biomass resources. Renewable energy sources can be used to generate heat or electricity, or can be used to produce liquid fuels for transport. For a general description of the various renewable energy sources see Box 1.

---

**Box 1**

**Sun:** The sun's energy can be used to generate power from photovoltaic cells, which convert light directly into electricity. Solar thermal energy is available by transferring energy from the sun via a liquid to heat water or air.

**Wind** energy is extracted by wind turbines, situated onshore or offshore. These transfer the momentum of passing air to rotor blades, which is then converted into electricity.

**Biomass** is a general term for material derived from growing plants or from animal manure. Biomass includes waste materials such as straw, or crops specifically grown as biomass fuel. Energy can be recovered through combustion of solid material, or of gases generated from the anaerobic fermentation of liquid material.

**Water** can be exploited as a renewable resource in a number of ways. Hydroelectric power is generated from the potential energy of inland water resources such as streams, rivers and lakes. Tidal energy is produced by the rotational energies of the earth, moon and sun, and can be used to generate electrical energy. Electricity can also be generated from waves. Waves receive their energy from the wind, which is mainly driven by the effects of solar heating in the atmosphere.

**Geothermal:** Some definitions include geothermal energy as a renewable energy source. Geothermal energy is the heat energy contained in rocks beneath the earth's surface and can be exploited to produce heat or generate electricity.

---

Renewable energy plays an important role in the process of integrating the environment into energy policy, through its potential to contribute to the objectives of sustainable development.

At the point of generation, renewable energy sources generally emit no greenhouse gases, with the notable exception of biomass, which is neutral over its life cycle in greenhouse gas terms. They also produce significantly lower levels of environmental air pollutants than fossil sources. Each EU Member State has committed itself to national targets to reduce or limit greenhouse gas emissions in order for the EU as a whole to meet the Kyoto Protocol obligation of reducing greenhouse gas emissions by 8 % from 1990 levels by 2008–12. Individual governments are working towards appropriate measures to achieve their own targets. The energy sector is one of the major emitters of carbon dioxide ($CO_2$) and other greenhouse gases, so increasing the use of renewable energy in place of fossil fuels can contribute towards achieving these targets.

Negative environmental impacts from renewable energy are, in general, lower than those encountered from fossil or nuclear energy sources and are usually more significant at the local level, near to the plant. Renewable energy installations may cause visual, noise or interference impacts, especially at the local level, although generally these can be minimised if the installation is planned and sited sensitively. The use of biomass necessitates additional transport. Biomass combustion also generates potentially polluting emissions, which need to be carefully regulated. The environmental consequences of growing biomass crops may include an impact from the use of agrochemicals, changes to water use, as well as changes to biodiversity and habitat, and visual alterations to the landscape. Hydro installations, in particular large installations, can have a local impact through construction activities, but will also affect water quality and flow, with consequent potential impact on the aquatic ecosystem. Photovoltaic (PV) systems have few effects when in operation, but the manufacture of PV cells needs to be carefully controlled due to the use of potentially toxic or hazardous materials.

Renewable energy can make an important contribution to security and diversity of energy supply, by providing a secure, indigenous source of energy that is available in a variety of forms to all Member States.

Renewable energy use is also important for reasons of social and economic cohesion. Renewable energy technologies are generally situated in regions with lower levels of investment or employment, such as rural or remote regions. Investment in new renewables plants in such areas can therefore benefit the local and regional economy.

The renewable energy industry is one of Europe's fastest-growing sectors, as Member States develop manufacturing capabilities to meet the growing demand for renewable energy, both domestically in the EU, and increasingly worldwide [4]. Building on a strong domestic market, European renewable energy companies already lead the world in their skills and expertise.

## 2.2.   Targets for renewable energy use in the EU

At the EU level, the European Commission's 1997 White Paper on renewable energy sources (European Commission, 1997a) set out the objective of increasing the share of renewable energies to 12 % of gross inland energy consumption [5] by 2010.

---

[4]   A recent EU study estimated that the renewable energy industry may generate up to 500 000 jobs by 2020 (ECOTEC, 1999).

[5]   The gross inland energy consumption is the total amount of energy which is consumed in an economy. See also Glossary.

| Figure 1 | Renewable energy sources as a contribution to gross inland energy consumption, EU |
|---|---|

**Source:** Eurostat.

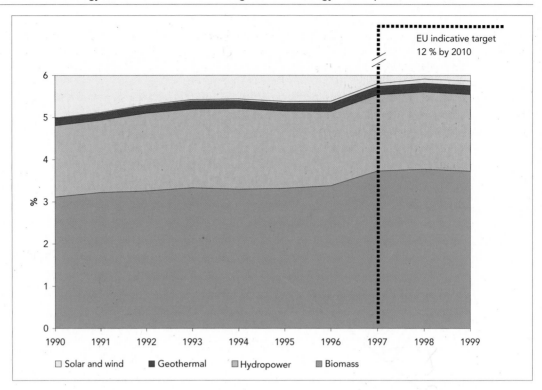

Renewable energy's share of gross inland energy consumption was 5.9 % in 1999, well short of the 12 % indicative target. Clearly the target is challenging for the EU. Indeed, even if the high growth rates observed between 1996 and 1999 are extrapolated, the share of renewable energies still falls short of the 2010 indicative target [6].

As a follow-up to the 1997 White Paper, in 1999 the Commission launched the Campaign for Take-Off (European Commission, 1999b). The campaign identified three key renewable energy sectors to be promoted during the period 1999–2003: solar energy (photovoltaics and solar thermal), wind and biomass (combined heat and power biomass installation, dwellings heated by biomass, biogas installations and biofuels). These also correspond to those technologies or market sectors where rapid uptake of renewable energy use is more likely to occur over this short timescale (to 2003), since they are already being successfully implemented in a number of Member States.

### 2.2.1. Renewable energy and liberalisation of the energy markets in the EU

The ongoing process of liberalisation of the electricity and gas markets in Europe has significant implications for renewable energy, particularly the proportion of electricity produced from renewable sources in the EU.

As a result of the 1996 EU directive on the internal market for electricity [7], Member States have been obliged to gradually open up an important proportion of their national electricity markets to competition which, combined with some other developments, has led to reduced prices for power from conventional sources.

However, despite many technological advances in the past decade, electricity from renewable sources is still more expensive than equivalent conventional power. This is mainly due to the relatively smaller size of renewable energy plant (which therefore cannot benefit from economies of scale) and the fact that external costs of fossil fuels have often not been fully internalised to level out the playing field between fossil and renewable energy. As the EU market becomes increasingly liberalised, it may become progressively more difficult for

---

(6)   This assumes that gross inland energy consumption will grow at the levels forecast by the Commission in its 1999 Primes baseline scenario (European Commission, 1999a).

(7)   Directive 96/92/EC of the European Parliament and of the Council of 19 December 1996 concerning common rules for the internal market in electricity.

renewable energy to compete in these changing markets without some support mechanisms over the medium term.

The recently adopted Directive to promote renewable electricity in the EU ([8]) attempts to address these issues. It aims to create a framework for electricity from renewables which will contribute towards achieving the indicative target of a 12 % renewables share in gross inland energy consumption (i.e. the total amount of energy consumed in an economy, including both heat and electricity) by 2010. The Directive requires Member States to take appropriate steps to encourage greater consumption of electricity produced from renewable energy sources by setting and achieving annual national indicative targets consistent with the Directive and national Kyoto commitments. The national indicative targets should be consistent with an increase in the share of electricity generated from renewable energy .sources in the EU from 13.9 % in 1997 to 22.1 % by 2010. Such national indicative targets, as well as the 1999 levels of renewable electricity share in gross electricity consumption ([9]) are shown in Figure 2.

| Renewable energy contribution to gross electricity consumption, 1999 | Figure 2 |

**Source:** Eurostat.

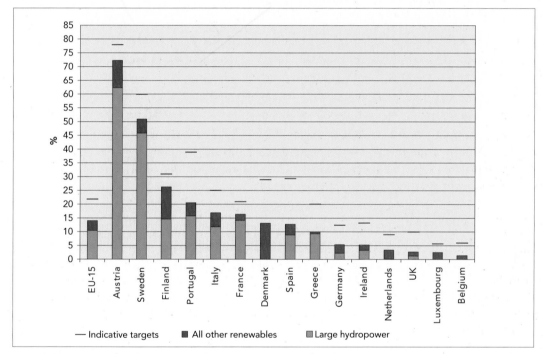

Notes:
1. Industrial and municipal waste has been included in 'All other renewables'. According to the renewable electricity directive (2001/77/EC) only the biodegradable part of industrial and municipal waste is renewable. Data on industrial and municipal waste presented here include all industrial and municipal waste, as there are no data available on the biodegradable part only. The share of renewable electricity in gross electricity consumption is therefore overestimated by an amount equivalent to the electricity from non-renewable industrial and municipal waste.
2. National indicative targets shown here represent reference values that Member States agreed to take into account when setting their indicative targets by October 2002, according to the renewables electricity directive (2001/77/EC).

In 1999 renewable sources contributed 14.2 % of gross electricity consumption: 10.5 % from large hydropower installations (a capacity of 10 MW or more) and 3.7 % from all other renewables.

Due to site limitations, the contribution of electricity from large hydro is expected to remain approximately constant between 1999 and 2010. Between 1996 and 1999 electricity from non-large hydro renewables (i.e. all other renewables except large hydro) grew rapidly. However,

---

(8)  Directive of the European Parliament and of the Council on the promotion of electricity from renewable energy sources in the internal electricity market (2001/77/EC).
(9)  See glossary.

non–large hydro renewables will have to grow significantly faster if the 2010 target is to be reached.

## 2.3.  Technologies covered under this study

This study concentrates on the technologies promoted under the Campaign for Take-Off.

For the purposes of the study the following renewable energy sources have been chosen:

- solar photovoltaics
- solar thermal
- wind
- biomass energy (biomass power including combined heat and power, biomass district heating and biofuels).

**Table 1**  **Renewable energy sources/technologies studied**

**Source:** Eurostat.

| Renewable energy source/ technology | Description | Units of measurement used in this study |
|---|---|---|
| Solar photovoltaics | Power generated using photovoltaic cells to convert light directly into electricity. | Power output (Gigawatt hours, GWh) |
| Solar thermal | Transfer of energy from the sun via a liquid to heat water or air. | Heat output (thousand tonnes of oil equivalent, ktoe) |
| Wind | Wind turbines extract energy from the wind by transferring the momentum of passing air to rotor blades, which is then converted into electricity. | Power output (GWh) |
| Biomass | A range of biomass fuels such as forestry and agricultural residues, and energy crops can be used to generate electricity in *power stations* including in *combined heat and power* plant, or can be used to produce heat in *district heating* plant. In addition, biomass can be used to produce transport fuels (*liquid biofuels*), primarily biodiesel and bioethanol, from processed agricultural crops and other biomass feedstocks. | Power output (GWh) or energy output as heat or fuel (ktoe) |

In 1999, the renewable technologies covered under this study together represented approximately 23 % of the non-hydro ([10]) renewables gross inland consumption (both heat and electricity) and 64 % of the non-hydro renewables gross electricity consumption. The remaining renewable energy sources not covered in this study are mainly waste combustion, including industrial and municipal waste ([11]) as well as contributions from other biomass sources and from geothermal energy.

Energy from biogas is included in the Campaign for Take-Off but is not considered in this study. Biogas generation is derived from the safe disposal of animal manure residues. Its exploitation is therefore dependent to a great extent on animal husbandry and waste management legislation. For this reason, the technology has not been studied further here.

Offshore wind energy is at an early stage of implementation, and so far there are only a few offshore plants in operation. This is why the penetration of offshore wind energy has not been included. Nevertheless Annex 4 provides a short analysis of lessons that could be learnt from previous experiences of renewable energy developments onshore, and in particular from the case studies presented here.

---

([10]) Hydro power is either large hydro power (over 10 MW) or small hydro power (less than 10 MW). The vast majority of hydro power in the EU is large hydro power. The contribution of large hydro power is expected to remain approximately constant in the coming decades due to site limitations. Hydro power (both large and small) is a site-specific resource dependent on water availability and thus has a number of different characteristics from those of other resources (such as wind and solar) which can be developed at a variety of sites.

([11]) In general, industrial and municipal waste also contains waste from non-renewable sources.

# 3. Identifying successful Member State/ technology combinations

**Photo:** Volker Quaschning

## 3.1. Selection criteria for identifying successful Member State/technology combinations

In order to identify those Member State/technology combinations where there has been most success in achieving renewable energy penetration, the following two selection criteria have been applied. This approach was carried out in each Member State and for each technology covered in this study, over the six-year period 1993–99.

- An absolute increase equivalent to at least 10 % of the total EU-wide increase in renewable energy output for that particular technology over the period 1993–99. The 10 % threshold was selected to identify those Member States which made the greatest contribution to the increase in renewable energy output of each technology in the EU.
- A percentage increase of renewable energy output of the examined technology greater than the EU-wide percentage increase for that technology between 1993 and 1999. This compares the percentage increase of each renewable energy source with the EU-wide percentage increase, and identifies those Member State/technology combinations which exceed the EU-wide figure.

This two-stage approach gives an opportunity to identify those Member State/technology combinations where exploitation of the technology is already well established, and continues to expand. It also highlights those Member States where a technology may be in its initial stages of take-up, with a rapid rate of increase in penetration but still with only limited quantities of energy output.

For the six-year period 1993–99, Eurostat data provides comprehensive statistics for all Member States and all energy sources. Although data for many Member States and energy sources are available for 1990–93, some of the annual data in this earlier dataset are not comprehensive. This period also pre-dates the time of greatest activity for renewable energy exploitation for most of the energy sources and Member States. This study therefore uses the 1993–99 dataset to identify the levels of penetration of each renewable energy source in each Member State.

## 3.2. Successful Member State/technology combinations and country-specific trends

The results of applying these two criteria are summarised in Table 2. This shows which Member State/technology combinations meet either of these two criteria. Background data relating to Table 2 are shown in Table 3. Figures 3–8 illustrate graphically the data presented in Table 3, by technology.

| Table 2 | Application of selection criteria to identify successful Member State/technology combinations for the period 1993–99 |
|---|---|

Source: Eurostat.

| Technology: Selection criteria (see note 1): | Photo-voltaics | | Solar thermal | | Wind | | Biomass: power | | Biomass: district heating (1993–98) | | Biomass: biofuels (see note 2) |
|---|---|---|---|---|---|---|---|---|---|---|---|
| Austria | | | ✓ | ✓ | | | ✓ | | ✓ | ✓ | ✓ |
| Belgium | | | | | | | ✓ | | | | |
| Denmark | | | ✓ | | ✓ | | ✓ | | | | |
| Finland | | | ✓ | | | ✓ | ✓ | | | | |
| France | | | | | | | ✓ | | | ✓ | ✓ |
| Germany | ✓ | ✓ | ✓ | ✓ | ✓ | ✓ | ✓ | | | | ✓ |
| Greece | | | ✓ | | | | | | | | |
| Ireland | | | ✓ | | ✓ | | | | | | |
| Italy | | | ✓ | | ✓ | | ✓ | | ✓ | | ✓ |
| Luxembourg | | | | | | | | | | | |
| Netherlands | | ✓ | ✓ | | | | | | | | |
| Portugal | | | | | ✓ | | | | | | |
| Spain | ✓ | ✓ | | | ✓ | ✓ | ✓ | | | | |
| Sweden | | | | | | ✓ | ✓ | | ✓ | ✓ | |
| UK | | | ✓ | | | | | | | | |

A tick (✓) indicates that the selection criterion was met, except in the case of biofuels (note 2 below).
Biomass power includes combined heat and power and refers to electricity output only; biomass district heating refers to heat output from heat plants only.
Note 1: Two criteria for selection are used:
    ✓ (left) represents a contribution of at least 10 % of the total EU increase in absolute terms, 1993–99;
    ✓ (right) represents a percentage increase greater than the EU percentage increase, 1993–99.
Note 2: Biofuels only:
    ✓ represents those Member States which indicate that they use biofuels (most do not).

| Application of selection criteria to identify successful Member State/technology combinations, 1993–99 Background data relating to Table 2 | | | | | | | | | | Table 3 |

| | Photovoltaics | | Solar thermal | | Wind | | Biomass power | | Biomass | |
|---|---|---|---|---|---|---|---|---|---|---|
| | | | | | | | | | district heating | |
| | 1993–99 | | 1993–99 | | 1993–99 | | 1993–99 | | 1993–98 | |
| | Absolute increase (GWh) | % increase | Absolute increase (ktoe) | % increase | Absolute increase (GWh) | % increase | Absolute increase (GWh) | % increase | Absolute increase (ktoe) | % increase |
| Austria | 1.4 | 233 | 37.1 | 150 | 42.0 | | 574.9 | 59 | 47.5 | 62 |
| Belgium | 0 | | 0 | | 4.9 | 60 | 140.3 | 238 | 0 | |
| Denmark | 0 | | 3.7 | 100 | 1 994.8 | 193 | 316.0 | 189 | -15.3 | -8 |
| Finland | 0 | | 0.2 | 8 270 | 45.0 | 1 125 | 2 697.0 | 47 | 20.2 | 33 |
| France | 0 | | 4.2 | 30 | 32.5 | 928 | 223.0 | 18 | 9.0 | 113 |
| Germany | 32.0 | 1 070 | 54.4 | 260 | 4 854.0 | 720 | 258.0 | 61 | 0 | |
| Greece | 0 | | 31.9 | 34 | 114.5 | 241 | 0 | | 0 | |
| Ireland | 0 | | 0.1 | 90 | 171.9 | 1 138 | 0 | | 0 | |
| Italy | 5.3 | 50 | 3.7 | 70 | 398.6 | 9 059 | 241.3 | 812 | 3.2 | 58 |
| Luxembourg | 0 | | 0 | | 18.0 | | 0 | | 0 | |
| Netherlands | 5.3 | 757 | 4.0 | 140 | 471.0 | 270 | 0 | | 0 | |
| Portugal | 0.5 | 100 | 2.9 | 20 | 112.0 | 1 018 | 179.0 | 20 | 0 | |
| Spain | 15.8 | 1 330 | 6.5 | 30 | 2 628.0 | 2 266 | 432.0 | 89 | 0.5 | |
| Sweden | 0 | | 0.6 | 10 | 319.3 | 618 | 898.0 | 42 | 205.4 | 56 |
| United Kingdom | 1.0 | | 4.8 | 90 | 678.0 | 310 | 0 | | 0 | |
| | | | | | | | | | | |
| EU increase 1993–99: | 61.3 | 358 | 154.1 | 65 | 11 884.5 | 502 | 5 959.5 | 50 | 270.5 | 39 |
| 10 % of EU increase 1993–99: | 6.1 | | 15.4 | | 1 188.5 | | 596.0 | | 27.1 | |
| Criteria thresholds used: | Higher than 6.1 GWh | Higher than 358 % | Higher than 15.4 ktoe | Higher than 65 % | Higher than 1 189 GWh | Higher than 502 % | Higher than 596 GWh | Higher than 50 % | Higher than 27 ktoe | Higher than 39 % |

Data in shaded cells meet the selection criteria, that is:

- at least 10 % of the total EU increase;
- a percentage increase greater than that for the EU.

Blank cells indicate that the value for 1993 is equal to zero.

Biomass in power includes combined heat and power and refers to electricity output only;
biomass district heating refers to heat output from heat plants only, and data refer to the period 1993–98.

**Source:** Eurostat.

| Figure 3 | Photovoltaics: absolute and % increase in generation, 1993–99 |

**Source:** Eurostat.

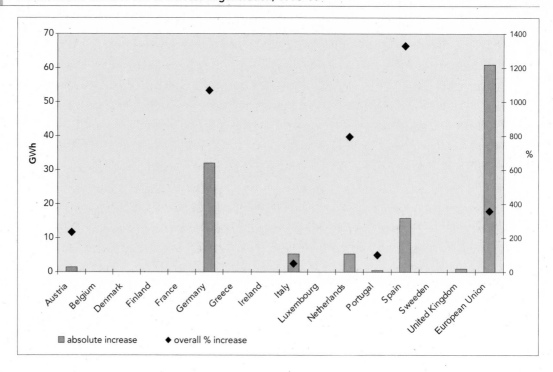

| Figure 4 | Solar thermal: absolute and % increase in generation, 1993–99 |

**Source:** Eurostat.

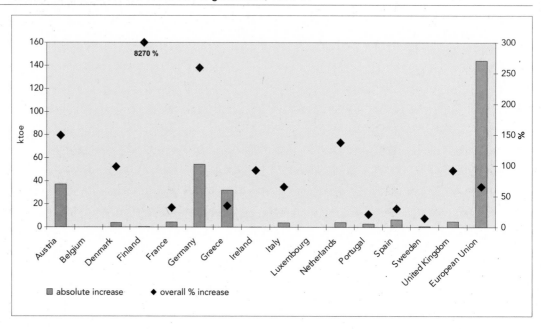

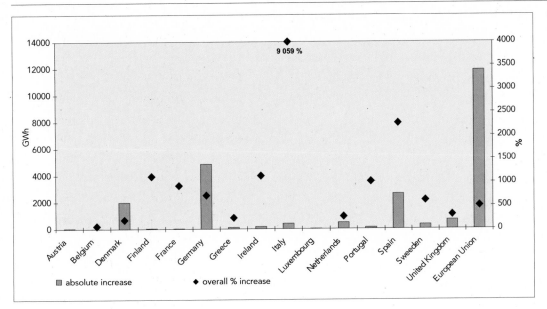

Wind: absolute and % increase in generation, 1993–99

Figure 5

**Source:** Eurostat.

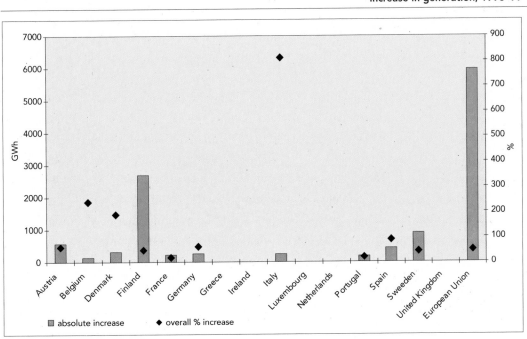

Biomass power generation and combined heat and power: absolute and percentage increase in generation, 1993–99

Figure 6

**Source:** Eurostat.

**Source:** Eurostat.

**Figure 7** · Biomass district heating: absolute and % increase in generation, 1993–98

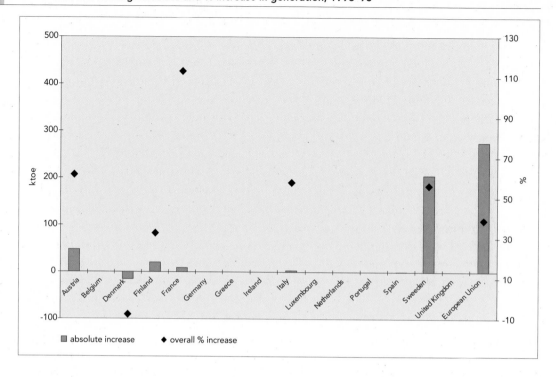

**Figure 8** · Liquid biofuels production in the EU, 1993–99

**Source:** Eurostat.

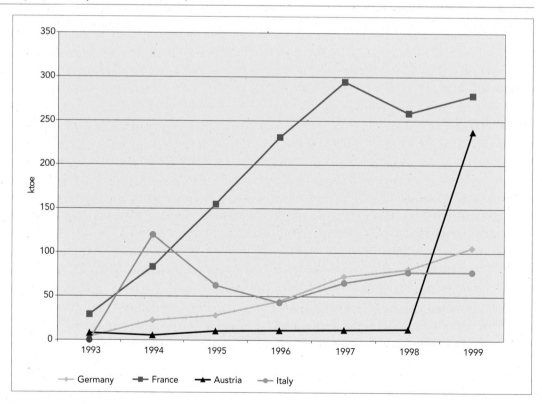

Table 2, Table 3 and Figures 3–8 show that the overall increase in renewable energy penetration achieved in the EU is not shared equally among the 15 Member States.

In terms of the absolute increase in renewable energy penetration, for most of the technologies only a few Member States contributed more than 10 % (each) of the total new resource output for the EU over the six-year period 1993–99[12]:

— two Member States (Germany and Spain) contributed 78 % of the new total EU output from photovoltaics;
— three Member States (Austria, Germany and Greece) contributed 80 % of new solar thermal installations;
— three Member States (Denmark, Germany and Spain) contributed 80 % of new wind output;
— two Member States (Finland and Sweden) contributed 60 % of new generation from biomass fuelled power stations (including biomass combined heat and power stations);
— two Member States (Austria and Sweden) dominated the increase in output from biomass district heating installations. Denmark's policy of replacing heat-only biomass district heating plants with biomass combined heat and power generation was evident from its reduced output over the period 1993–98;
— only four Member States use biofuels to any significant extent. France is the market leader, producing about 40 % of the total.

When the second selection criterion is applied, a greater number of Member States achieved a percentage increase in output higher than the percentage figure for the EU as a whole over the period 1993–99[13]. Three Member States surpassed the EU percentage figure for photovoltaics. For solar thermal and wind, eight Member States exceeded the EU figure and for biomass in power stations (including combined heat and power stations) six Member States exceeded the EU figure over this period. The data for biomass district heating show that over the period 1993–98, only four Member States achieved percentage increases higher than that for the total EU.

Only a few Member State/technology combinations recorded positives for both criteria, i.e. a rapid **and** a significant increase in renewable energy output over the period (see Tables 2 and 3). Germany achieved the greatest levels of new renewable penetration over the period and met both of the criteria for all the technologies except biomass. Positive combinations are also highlighted in some other Member States — in Spain (photovoltaics and wind), Sweden (biomass district heating) and Austria (solar thermal and biomass district heating).

Many Member States show either a large absolute increase in renewable energy output for individual technologies, **or** a large percentage increase. The fact that only a small number of Member State/technology combinations meet both of the criteria usually relates to their starting level of renewable energy output for the technology in 1993. Thus Member States with only very low initial levels of renewable energy use may demonstrate rapid growth rates even though the actual quantity of output is still only small. Examples here include Finland, which increased its solar thermal output over the period by 8 270 %, but in absolute terms output rose by a very small quantity (0.2 ktoe).

Conversely, Member States with levels of renewable energy use that were already relatively high in 1993 show a less rapid percentage growth rate, but these Member States may still have added significant quantities to the overall output. For example, Denmark's wind industry was already well established by 1993, but it added a further 1 995 GWh by 1999, an increase of 193 %. In Finland, power from biomass was already high in 1993, but continued to expand steadily, adding 2 697 GWh over the period, a 47 % increase over the 1993 level.

One Member State (Luxembourg) did not meet either of the two criteria for any of the technologies, while Belgium, Greece, Portugal and the UK met one of the criteria only once.

---

(12) For biomass district heating, the period examined was 1993–98.
(13) Idem.

# 4. Potential factors for success

**Photo:** Volker Quaschning

## 4.1. Barriers to success

Successful penetration of renewable energy can only be achieved after overcoming many varied obstacles to an increase in their exploitation.

A large amount of information and research has already been carried out at both the EU and Member States levels focused on identifying and breaking through these barriers. Much of the work at European level was drawn together through a series of consultations to set out the framework for future initiatives and a strategic action plan for overcoming the barriers to renewable energy implementation (European Commission, 1997b). The EU White Paper on renewable energy sources (see Section 2.2) also summarises these barriers and the measures to be taken to counter them.

Renewable energy projects are generally of a smaller scale than conventional energy projects and consequently cannot benefit from economies of scale to the same extent. In relative terms, they also have high capital costs which need guarantees of long-term stable income streams to ensure financial viability. It is therefore important that the non-technical frameworks in place do not discriminate against these kinds of projects but enable them to be brought forward as attractive and financially viable schemes.

Table 4 summarises the main types of barrier that have been identified together with the general obstacles associated with them.

| Barriers and obstacles to renewable energy deployment | Table 4 |
|---|---|

| Barrier | Obstacle |
|---|---|
| Political | Lack of political motivation to support the market initiatives needed for the development of renewables |
| Legislative | Lack of an appropriate legal framework and legislation at EU and national levels that support the development of renewables<br>Difficulties with linking electricity or heat from renewables into the existing electricity and heat networks |
| Financial | Lack of appropriate financing for long-term financial benefits |
| Fiscal | Renewable energy technologies suffer from distorted competition from conventional energy sources (e.g. coal, nuclear) in terms of final end-user prices |
| Administrative | Lack of practical support at the regional and local level to stimulate development of renewable energy projects |
| Technological | Technological obstacles related to research, development and demonstration |
| Information, education and training | Lack of awareness of the potential and possibilities for renewables |

The expansion in the use of renewable energy during the 1990s demonstrates that, for some technologies in some Member States, there are factors that appear to be acting in favour of implementation. Nevertheless, as shown in Table 2 and Table 3 in the previous section, it is apparent that for other technologies and in other Member States, the level of implementation is not as rapid, and obstacles are still preventing implementation.

Clearly, the situations in individual Member States and for individual technologies have created different framework conditions, some of which are more favourable towards increased exploitation of renewables than others. However, in many instances it can be difficult to identify the conditions that have resulted in successful implementation of projects — or it may be difficult to replicate these conditions in new market conditions.

## 4.2.  Factors influencing renewable energy deployment

This study aims to identify the range of factors that can influence the likely successful implementation of renewable energy projects and assess how they work.

The principal factors studied are based around the key criteria given in Table 4.

- **Political**: How strong is political support for renewable energy?
- **Legislative**: How accessible is the energy market to independent electricity producers?
- **Financial**: How accessible is funding for investments in renewable energy projects?
- **Fiscal**: How favourable is the fiscal infrastructure for renewable energy?
- **Administration**: How favourable are the administrative arrangements for obtaining permission to construct a renewable energy project?
- **Technological development**: Is there support for the development of strong national capabilities in renewable energy technologies?
- **Information, education and training**: Is support given to widely disseminate information on the benefits of renewable energy?

Each factor is discussed in more detail below. These factors will be used in the following section to evaluate those Member State/technology combinations that met both the selection criteria for successful penetration. They will also be used in Annex 4 to evaluate the most interesting cases of those Member State/technology combinations that met only one criterion.

### 4.2.1. Political

*How strong is political support for renewable energy?*

EU-wide support for a greater use of energy from renewables is demonstrated strongly through the 1997 EU White Paper on renewable energy sources, and more recently the 1999 Campaign for Take-Off and the 2001 directive on renewable energy in the internal electricity market. All of these form the basis for action plans for subsequent implementation at Member States level, through coordinated activities at national, regional and local levels. In particular, support is demonstrated through the adoption of **policies in favour of renewables**. These are often developed in conjunction with the development of an overall **national energy plan**. One of the most important elements to consider here is the adoption of **official targets for the level of uptake of renewables**, usually expressed as a target level of use of renewable energy (heat and/or power) in the Member State. For Member States that have strong regional governments, these national targets may also be translated to appropriate regional targets. Alternatively, there may be no national targets at all, but only targets developed at the regional level.

### 4.2.2. Legislative

*How accessible is the energy market to independent electricity producers?*

For small independent power generators (renewable or fossil-fuelled), it is essential that there are access agreements that permit them **access to the electricity grid** to distribute their electricity. The costs and logistics of connecting to the grid can be significant for small energy schemes, and the imperfections associated with grid connection arrangements have not yet been removed in many Member States.

Another important aspect for renewable power producers is the availability of a **market** for their power, usually achieved by connection and transmission via the main electricity supply grid. Due to the generation characteristics of renewable energies and their higher generation costs compared with other forms of energy, the absence of a guaranteed market for renewable power seriously jeopardises the financial viability of many renewable energy projects.

Different Member States have adopted different approaches to providing support for a guaranteed market for renewable generation, but generally there have been two main mechanisms which governments have used:

**'Feed-in'** arrangements — with a fixed price for electricity generated, and purchase/sale arrangements available continually, coupled with a purchase obligation by the utilities. The feed-in arrangement is a legal obligation on utilities to purchase electricity from renewable sources. It also sets the prices to be paid for electricity generated from these renewable sources, generally higher than the price that would be available outside the tariff mechanism. These tariffs may vary for different renewable energy technologies, and are generally linked to prices paid by final consumers.

**Tendering** arrangements — a competitive process based on periodic calls for tender, to support a certain predetermined quantity of renewable electricity. Under this system, a fixed amount of renewable energy is supported through a levy on electricity consumers, which is directed at specific renewable technologies through a tendering system. This system enables specific renewable technologies to be supported, even those at differing stages of technological development, because the tendering system encourages competition within technologies. This has resulted in cost reductions for many of the technologies supported.

There are two more recent developments in new initiatives to support electricity from renewable sources:

**Green pricing schemes.** Green pricing allows consumers to support the generation of electricity from renewable sources by paying a voluntary premium. The schemes being developed are varied in their design, ranging from consumers paying a fixed premium per year to support new renewables developments (whilst themselves still receiving some 'brown' or non-renewable power), to receiving all of their power from renewable sources.

**Green certificates.** Green certificates benefit producers of renewable electricity and are issued according to the amount of renewable electricity produced or sold into the grid. Demand for green certificates can come from several sources. There may be voluntary demand from consumers who wish to purchase 'green electricity'. The government can also stimulate demand by stipulating that suppliers should provide their consumers with an increasing proportion of their power from renewable sources. The value of the certificate can assume a separate market value through trading between electricity suppliers. Trading in green certificates is being developed in a number of Member States and at a European-wide level.

### 4.2.3. Financial
*How accessible is financial support for investments in renewable energy projects?*
The capital costs of building renewables projects are a significant barrier to their implementation, especially for newer technologies that are more costly and that have less of a track record in implementation (e.g. photovoltaics). Finance may be provided from either public or private sources:

**Public support** can be made available for renewable energy projects through grants or loans:

- grants — public sector support at EU, national, regional or local level;
- low-interest loans — usually through national or regional financial institutions with public subsidy support;
- loan guarantees — again, usually provided with public subsidy support.

**Private sector** funding of renewables projects from banks and other financial institutions such as venture capital organisations is of vital importance to the long-term commercialisation of renewables. It can be provided either in conjunction with or without public funds, depending on the financial viability of the project.

### 4.2.4. Fiscal (taxation)
*How favourable is the fiscal infrastructure for renewable energy?*
Energy produced from renewable sources currently competes against energy from conventional sources at a disadvantage in many cases. There are several reasons for this. Many of the fossil and nuclear fuel plants operating in Europe today were built with significant subsidies, while much of the large fossil generation is from older power plants where the capital costs have now been recovered. Both of these factors enable these types of plant to generate at lower cost compared with new plant which must take capital investment costs into account. This is particularly challenging for renewable energy plant because capital costs contribute a very high proportion of total plant costs.

In addition, the external costs of energy production from fossil or nuclear sources are sometimes not fully taken into account when deriving electricity generation costs. External costs from nuclear or fossil fuel plant include their contribution to pollution through carbon dioxide, sulphur dioxide, nitrogen oxides and other emissions and pollutants (fossil fuel plant), or to nuclear waste generation and to risks of radioactive contamination (nuclear plant). Benefits from renewable energy reflect the benefits of energy provision from non-fossil, clean energy sources, and the potential benefits of providing decentralised power production (embedded generation).

As a result of all these and other issues discussed in this section, costs of renewable generation are currently higher than that of conventional (fossil or nuclear) generation.

Some Member States have started to address these issues, usually through the imposition of a **carbon or energy tax**. These aim to modify energy consumption levels and patterns to encourage greater energy efficiency and a greater use of renewable resources. Some Member States also provide **tax exemptions/reductions** or **tax incentives** (for example accelerated depreciation of capital investments in renewable energy) to companies or individuals making use of or investing in renewables-related goods or services. These encourage investment in clean energy projects.

### 4.2.5. Administration

*What are the administrative arrangements for obtaining permission to construct a renewable energy project?*

Public authorities have an important role to play in supporting the conception, siting and development of new renewable energy projects. **Achieving planning permission** can be one of the biggest barriers to implementation of new renewables projects in some countries or regions. This is especially true for wind projects, but larger biomass and many hydro installations also encounter problems related to their construction. However, the ease with which planning permission can be obtained varies between different Member States. Responsibility for giving permission may be at the local, regional or national level, and may be carried out on a project-by-project basis, or may form part of a wider planning process that has already provided a framework for the location of new projects.

### 4.2.6. Technological development

*Is there support for technological development of strong national capabilities in renewable energy technologies?*

Support for research and development (R&D) plays a vital role in the progression from research and technological development through demonstration to final full-scale commercialisation of a new technology. All renewable energy technologies benefit from R&D support to ensure the continued development of a strong and competitive industry. Support is especially important where renewable energy technologies are still at early stages of development — for example photovoltaics.

Technological support focuses not only on R&D, but also on demonstration and implementation of new technologies as they mature. For a Member State to build up its indigenous capabilities in a developing market such as the renewables market, it is important for the emerging industry to be given consistent and targeted support for demonstration and implementation projects.

### 4.2.7. Information, education and training

*Is support given to widely disseminate information on the benefits of renewable energy?*

Dissemination activities to promote the benefits of renewables are a vital component of a national renewables support programme. These include the provision of information to all of the principal actors and groups involved with the development of renewables — financiers, planners, politicians (local and national), and the general public — to raise awareness and educate about the potential of renewable energy. It is also very important to support training initiatives for new renewable energy developments, such as for solar heating installers or for farmers wishing to grow new energy crops.

The development of renewable energy is closely linked to its level of acceptance by the people who will benefit from it and who will see the new renewable energy projects being built in their area. Public acceptability is vitally important for new renewable energy developments, particularly since projects are often smaller scale and their greatest impacts occur at the local (community) level.

# 5. Member state/technology examples of successful penetration

**Photo:** Volker Quaschning

Table 2 (Section 3) highlighted the principal Member State/technology combinations that appear to be successful, in terms of rate and/or the amount of increase in output over the six years 1993–99.

The preceding discussion has summarised the key factors that may influence the likely successful penetration of renewable energy technologies. These factors are now assessed in a series of case studies which reflect the successful Member State/technology combinations identified in Section 3. Where possible and appropriate these case studies give representative examples of renewable energy projects.

In this section, case studies are evaluated which demonstrate those Member State/technology combinations that meet **both** the **criteria** for successful penetration. These combinations are listed in Table 5:

| Member State/technology combinations meeting both the criteria for successful penetration, 1993–99 | | Table 5 |
|---|---|---|

| Country | Renewable energy technology | Applications |
|---|---|---|
| Austria | Biomass | District heating |
| | Solar thermal | Solar thermal collectors |
| Germany | Photovoltaics | Photovoltaics in urban areas |
| | Solar thermal | Heating initiatives in various cities |
| | Wind energy | Wind farms |
| Spain | Photovoltaics | Grid-connected photovoltaics |
| | Wind energy | Wind farms |
| Sweden | Biomass | District heating |

In Annex 1 the most interesting case studies of Member State/technology combinations that meet **one** of the **criteria** for successful penetration are evaluated. These combinations are listed in Table 6.

| Table 6 | Most interesting Member State/technology combinations meeting one of the criteria for successful penetration | | |
|---|---|---|---|
| | Country | Renewable energy technology | Applications |
| | Denmark | Biomass | Power |
| | | Wind energy | Wind cooperative |
| | Finland | Biomass | Combined heat and power |
| | France | Biomass (biofuels) | Biodiesel |
| | | Biomass | District heating |
| | | Wind energy | Wind developments |
| | Germany | Biomass | Power and district heating |
| | Greece | Solar thermal | Solar hot-water systems |
| | Ireland | Wind energy | Wind farm |
| | Italy | Wind energy | Wind farms |
| | Netherlands | Photovoltaics | Roof-integrated photovoltaics |
| | Portugal | Wind energy | Wind projects |
| | Spain | Biomass | Power |
| | Sweden | Biomass | Power |
| | | Wind energy | Wind farms |

## 5.1.  Austria — Biomass district heating

*Austria has extensive forestry and other biomass resources that are used as energy resources. Between 1993 and 1998, it achieved significant increases in its level and rate of use of biomass for heat production in general and especially for district heating purposes.*

| | |
|---|---|
| In 1993: | 77.1 ktoe |
| In 1998: | 124.6 ktoe |
| | |
| Increase 1993–98: | 47.5 ktoe, 62 % |

District heating is very common in Austria, and the use of biomass as fuel is increasing. By 2000 there were more than 500 district heating plants (totalling more than 650 MW installed capacity) operating throughout Austria.

**Success factors:**

• *Political: National and regional support to expand the use of biomass*

Austria has few indigenous fossil fuel resources, so its energy policy addresses security of supply issues through promoting energy efficiency and reducing use of imported fuels, combined with stimulating the use of renewable energy. Austria's large biomass resources play an important role in increasing the use of renewable energy. The government and, in particular, the regions provide active political support for biomass energy. Several regions have biomass-related targets.

• *Fiscal: Energy taxes favour renewable energy schemes*

Austria introduced an energy tax on the use of gas (EUR 0.0435/m³ (cubic metre) + 20 % VAT) and electricity (EUR 0.003/kWh + 20 % VAT) in 1996. The tax applies to small-scale as well as industrial users. Part of the tax revenue is made available to the *Länder* and to the communities for the implementation of energy saving and environmental protection measures, including measures to promote renewable energy.

• *Financial: Public grants and subsidies for biomass installations*

Support is provided at both the national and regional level for biomass installations, particularly for district heating schemes. Eligible regions have also benefited from EU Structural Funding support targeted at renewable energy schemes including biomass.

The support includes:

— subsidies of 10–30 % of eligible costs through a national environmental support programme;
— regional support plans that provide subsidies of up to one third of the costs;
— local and regional support targeted towards private households to subsidise the cost of connection to heating networks;
— special support programmes by the farmers' association to encourage farmers to invest in biomass plants.

• *Administration: Long history of public support for and use of biomass as a fuel resource*

Austria has a decentralised population structure and a densely wooded landscape, and has been using wood as an energy resource for centuries. Decentralised heat production from biomass is accepted and promoted by the local and regional authorities.

The local council implements planning decisions at the local level. There can be initial opposition to district heating proposals. In some cases where this occurred, public authorities took the lead and established connections to public buildings to demonstrate the benefits.

- *Technological development: Indigenous manufacturing expertise*

New technological developments for biomass production processes are supported both within Austrian universities and in association with industry. There is already a well-established local industry that developed to meet the demand for new biomass district heating plants, including boiler and pipework manufacture, and installation services.

- *Information, education and training: Long history of use of biomass as fuel, benefits to key local economic actors from biomass projects, promotion of benefits from biomass use by energy agencies*

Biomass use is very well established and accepted in Austria, both at the local level for small-scale applications and at the industrial level, due to the country's extensive wood-based industries. At the larger-scale and industrial level, farmers are supportive of new biomass projects because of the additional income that will be generated. Wood users such as sawmills also benefit because they have an additional market for their wood wastes. These actors, in particular the farmers, have been key in increasing public acceptance of biomass projects.

At the local level, most regions carry out active dissemination activities to promote the economic and environmental benefits from using biomass as a fuel by individuals or communities. These activities are usually coordinated through regional or local energy agencies, which place strong emphasis on the importance of institution building and on activities in the information, communications and training sectors. The overall result of these activities is that the general public is well informed about the benefits and use of renewables.

## 5.2. Austria — Solar thermal

*Austria's rapid increase in its use of solar energy for heating purposes demonstrates that solar thermal can provide an important source of energy even in less sunny regions.*

| | |
|---|---|
| In 1993: | 24.9 ktoe |
| In 1999: | 62.1 ktoe |
| Increase 1993–99: | 37.1 ktoe, 150 % |

One of the most successful regions to encourage solar thermal uptake is Upper Austria. In 1994, the Upper Austrian Energy Plan established targets to reduce fossil fuel consumption, improve energy efficiency and increase the use of renewable energy in the region. These targets, which were met, included reducing domestic energy use by 20 % and increasing the proportion of energy provided from renewable sources to 30 % by 2000. The Upper Austrian Energy Agency has been one of the main promoters of actions to achieve these targets, and has been very successful through a combination of support measures from both national and local sources, combined with a high level of information dissemination activities to raise awareness in the region.

More than 500 000 m² of thermal solar collectors had been installed in Upper Austria by 2000 as a result of the energy plan, for space heating and producing hot water, mainly in domestic buildings. By 2000 the region had about 0.4 m² installed solar collectors per inhabitant — one of the highest densities of solar collectors per inhabitant in Europe. Similar initiatives are now being implemented in other Austrian regions, supported through local or regional energy plans.

**Success factors:**

- *Political: National and regional support towards development of renewable energy*

The Austrian national government's energy policy places a strong emphasis on improving the country's security of energy supply and reducing the amount of energy imports. This, combined with a commitment to environmental protection, has resulted in a long-standing level of political support for renewable energy at the national level and has led to active political support at the regional level. The government also links the benefits of supporting renewable energy to associated socio-economic impacts of job creation and economic benefits to the local economy.

- *Financial: Loans and grants available through national and regional government*

Financial support for installing solar collectors is available through support programmes both at the national level (targeting companies) and at the regional level (targeting households). This support includes grants to householders and low-interest loans on investments in renewable energy. The high level of financial support available for installing solar collectors has been one of the main reasons for the high uptake of these systems in the Upper Austrian region. Other regions are now following the example of Upper Austria: for example, the town of Feldkirch in western Austria subsidises the installation of solar hot-water collectors by 25 %, while the Municipality of Graz provides EURO 35/m² solar hot-water collectors installed.

- *Fiscal: Energy taxes favour renewable energy schemes*

An energy tax was introduced in 1996 on both small-scale and industrial users of gas and electricity. Part of the revenue from this energy tax is recycled to support various environmental protection measures, including support for renewable energy.

- *Administration: Active local and regional support towards the installation of solar collectors*

Many of the regional and local administrations have encouraged the installation of solar collectors through active support measures, for example through installing collectors on their own municipal buildings.

- *Technological development: Support for indigenous solar collector manufacturers and installation industries*

Solar thermal benefited during the 1980s from the establishment of a network of groups throughout the country which provide active support and advice to individuals wishing to install solar water heaters themselves. This initiative encouraged the development of an industry to provide commercial installations. Austria now has a network of companies working in the field of solar collectors, as well as in other renewable energy technologies, which in Upper Austria alone employs more than 1 000 people.

- *Information, education and training: Positive dissemination and support for renewable energy use through the regional energy agencies*

Regional energy agencies provide expert advice, research and targeted initiatives aimed at promoting national and regional energy policies. In Upper Austria, the agency provides targeted support towards promoting the uptake of solar collectors through a wide-ranging programme. Initiatives include:

— information and training for specialised installers
— marketing activities
— demonstration systems to show successful installations.

## 5.3.  Germany — Photovoltaics

*Germany has the highest level of photovoltaics installations in Europe, and the third highest in the world, after the United States and Japan.*

| In 1993: | 3.0 GWh |
|---|---|
| In 1999: | 35.0 GWh |
| Increase 1993–99: | 32.0 GWh, 1 070 % |

Many of Germany's regions actively support photovoltaics (PV) as part of their efforts to expand the use of renewable energy. Berlin, for instance, is sometimes referred to as the solar capital because of the rapid increase in PV installations. More than 9 000 m$^2$ of PV modules have been installed in Berlin, with a total generating capacity of nearly 800 kW. These include the presidential residence, the town hall, and many other ministries and public buildings.

The new Innovation Centre for Environmental Technology in Berlin-Adlershof was developed by the Berlin Energy Agency, and financed through a partnership between the Energy Agency, the local energy provider BTB and the owner of the building. They jointly invested in a solution to cover the working expenses of the solar installation and to sell electricity into the energy grid.

This and many of the other installations in Berlin provide excellent examples of successful PV installations, and of ecological building management.

**Success factors:**

• *Political: National and regional support towards the development of photovoltaics*

German energy policy is closely linked with national policies to support climate protection. Renewable energy plays an important part in this policy, and the government has actively supported financial provision towards renewables, both at national and regional levels. Most German regions also have energy policies, targets and support mechanisms designed to encourage the development of renewable energy. PV especially has benefited from support from regional governments. For example, Berlin's energy policy was established in 1994 and amongst other things seeks to increase support for the use of renewable energy and reduce CO$_2$ emissions by 25 % per person from 1990 levels by 2010.

• *Legislative: Premium-set tariffs combined with an obligation to purchase provide a stable, commercially favourable market to renewable electricity producers*

The Electricity Feed-In Law supports renewable electricity, including PV, by providing a guaranteed market and fixed price for the electricity produced from renewable energy sources. From 1 April 2000 the tariffs for PV are even more advantageous: they have been raised sixfold from EUR 0.08/kWh (DEM 0.16/kWh) to EUR 0.51/kWh (DEM 0.99/kWh). This rate is proving very attractive: the German government had to limit applications for receiving this new tariff in 2000 because the amount of money set aside for supporting it had already been reached.

• *Financial: Loans and grants available for photovoltaic schemes*

An important stimulus to developing PV in Germany have been the PV roofs programmes. The 1 000 PV roofs programme started in 1991 and provided subsidies for production costs of PV units of 60 % in the new *Länder* and 50 % in the rest of Germany. The programme was successfully completed in 1996/97. A follow-on 100 000 PV roofs programme started in 1999, which provides EUR 560 million towards supporting individuals and small and medium-sized companies to install grid-connected PV schemes. The recent increase in tariffs available to PV installations and the corresponding rapid increase in the number of installations has meant that the target date for achieving the 100 000 roofs programme has been brought forward by one year to 2003.

The national European Recovery Programme (ERP) for Environment and Energy Savings offers long-term loans with low interest rates for investment in the use of renewable energy. The programme is administered by the federally owned Deutsche Ausgleichsbank (DtA), Bonn. Loans may amount to 50 % of investment costs, and provide favourable interest rates and loan arrangement. This programme provides grants for capital subsidies to households for PV installations.

- *Technological development: Germany has developed a successful domestic photovoltaics cells and components manufacturing industry*

High levels of federal support for research and development have helped to build a strong and competitive domestic PV industry. Almost one third of the funding support provided from the federal government towards energy research and technology is focused on renewable energy. In 1996 expenditure was approximately EUR 100 million. The majority of the funds are devoted to solar applications, particularly PV. This, combined with the expanding domestic PV market, has made Germany a location of choice for new manufacturing plants. More than half of Europe's PV manufacturing capacity is now located in Germany.

- *Information, education and training: Local energy agency stimulates private and public interest in photovoltaics*

At the local community level, there is high environmental awareness among German citizens, particularly of energy issues. In addition, energy agencies play an important role in stimulating the demand for PV installations.

In Berlin, for example, the approach is to encourage links between private sector and public organisations to build strong local partnerships to implement PV solutions. To achieve this, the Berlin authorities supported the establishment of an energy agency, in conjunction with developing support programmes. The Berlin Energy Agency is responsible for informally coordinating active links between the Berlin Senate, Berlin businesses and energy utilities to implement new PV projects on new public and private buildings.

## 5.4.  Germany — Solar thermal

*Germany is leading the way on solar thermal energy.*

| | |
|---|---|
| In 1993: | 21.0 ktoe |
| In 1999: | 75.4 ktoe |
| Increase 1993–99: | 54.4 ktoe, 260 % |

Many German municipalities have seen a rapid increase in the uptake of thermal solar installations. Examples of some successful municipalities in south-west Germany include:

— Freiburg: more than 200 domestic installations (2 500 m²) achieved by 1996;
— Friedrichshafen: nearly 5 000 m² of collectors installed to support solar-assisted small-scale district heating and hot-water systems;
— Ulm: installation of a central hot-water system for 38 homes and a district heating system (heat and hot water) for 86 residential units, in which heat from solar collectors is used together with a combined heat and power plant fired by biomass and gas.

These examples illustrate the variety of innovative ways in which German municipalities are promoting a greater use of solar energy for thermal energy requirements. Both small and larger-scale installations (more than 500 m²) are being established, and interest in large-scale solar heating opportunities is increasing, with most of these installations supplying heat to residential buildings.

**Success factors:**

- *Political: National and regional support towards increasing use of solar thermal installations*

Political support from national and regional governments is translated into practical implementation measures in the form of targets, grants, research support and other actions aimed at increasing the level of renewable energy use. Many regions or municipalities have targets for increasing their level of renewable energy use, which often place a strong emphasis on renewable energy and energy savings, linked with climate protection objectives. Regions also see solar hot-water systems as important components in implementing Local Agenda 21 (opportunities for carrying out climate protection initiatives at the local level) in their region.

- *Financial: Federal government and private sector financial support to solar thermal installations*

A wide range of federal, regional and private sector financial support is available for solar installations. These include:

— Federal government support: The federal government's Solarthermie 2000 demonstration programme subsidised the construction of the long-term hot-water storage and the district heating system in Friedrichshafen, contributing 53 % towards total costs.
— Regional support: In Friedrichshafen, 9 % of total costs came from regional support measures.
— Energy utilities: The role of the local municipal energy company is also important. In Freiburg, for example, the company provided financial support of EUR 230/m² (part-financed from EU funds). In Friedrichshafen, the owners and operators of the district heating system have responsibility for overall risk and guarantees of the system.

Local people can also benefit from low interest rates provided by local or regional banks for solar installations. A number of financial institutions in Germany have supported both community- and non-community-based renewable energy projects with favourable financing packages.

- *Administration: Active support provided from municipalities for solar thermal installations*

One important factor is the role played by both the municipality and the local energy utility in encouraging the uptake of solar heating. The municipality collaborates closely with the utility in establishing and implementing energy planning and municipal energy policy and targets.

- *Technological development: Quality standards provided to guarantee results from solar collectors*

GRS (Guaranteed Results from Solar collectors) is an initiative established by several municipal utilities in 1993, coordinated by municipalities. Its purpose was to ensure that the annual amount of heat supplied by a solar collector was guaranteed by the manufacturer, thus allowing for precise cost–benefit calculations and ensuring that systems met customer needs. Its application has helped to boost consumer confidence in the quality of the new types of heating systems available in Germany, and in particular those produced by German manufacturers.

- *Information, education and training: Active promotion of the benefits of solar thermal installations by municipalities, utilities and local energy agencies*

Close collaboration between municipality and utility has led to a combined effort to both disseminate and improve public relations in order to promote renewable energy uptake. Specialist advice centres have been established through local energy agencies to provide technical and practical support to local associations for solar installations, but also to support demonstration schemes.

Public awareness of energy and environmental problems, and the opportunities for renewable energy, is strong in Germany.

## 5.5.  Germany — Wind energy

*Germany has established itself as the world leader in wind power, with the help of a feed-in law.*

| In 1993: | 674.0 GWh |
|---|---|
| In 1999: | 5 528.0 GWh |
| Increase 1993–99: | 4 854.0 GWh, 720 % |

Germany increased the number of turbines installed considerably during the 1990s. Many of the projects established in Germany before the mid-1990s were small to medium-sized installations, often with a high degree of participation from the local community. One example is the Halde Nierchen wind farm in the state of Nordrhein-Westphalia. This wind farm comprises nine 1-MW turbines. The wind farm was built in 1998, and all power generated from it is sold to the local public utility EBV under the Electricity Feed-In Law.

**Success factors:**

• *Political: National and regional support towards wind energy development*

Federal support for renewable energy started more than 10 years ago with the 250 MW Wind Programme. This was initiated in June 1989 as a 100 MW Wind Programme and was extended in February 1991 to 250 MW. The programme's aim was rapidly achieved, and by 1999 Germany had more than 4 000 MW installed capacity. Political support from government is mainly through the feed-in law, grants and research support. Many German regions or municipalities have targets for increasing their level of renewable energy utilisation, including wind energy use.

• *Legislative: Premium-set tariffs combined with an obligation to purchase provide a stable, commercially favourable market for renewable electricity producers*

The single most important factor for the rapid and successful implementation of wind energy in Germany is its feed-in law. From 1991–2000, the Electricity Feed-In Law provided a guaranteed market and fixed price for the electricity produced from renewable energy sources. Wind energy schemes benefited considerably during the late 1990s from the favourable tariffs available through these support measures, and the feed-in law and its successor, the Renewable Energy Sources Act (2000), continue to be the principal mechanism for achieving the rapid uptake of wind energy in Germany.

Under the feed-in law, operators of the grid were obliged to purchase electricity produced from renewables within their respective supply areas, at agreed and fixed prices. For wind power the price available in 1997 was DEM 0.1715/kWh (EUR 0.088/kWh). In order not to overburden grid operators in areas where there were high rates of renewables generation with having to purchase at premium prices, a limit of 5 % renewable electricity was set from 1998 that applied within each region. Above this mark, operators of the grid were exempted from the obligations of purchase and refund.

As the amount of electricity from renewable sources expanded, a number of regions exceeded the 5 % ceiling. There was also an uneven financial burden between grid operators in regions close to the 5 % ceiling and regions with low levels of renewables generation. In order to address these and a number of other issues the law was replaced by the Renewable Energy Sources Act (2000). This continues to provide a guaranteed market and fixed favourable tariffs for electricity generated from renewable sources. The 2000 Act abolished the 5 % cap and introduced a system that allows transmission grid operators to share amongst themselves the costs of compensation to renewable electricity producers.

• *Fiscal: Tax exemptions are available to investment in renewable energy technologies*

Private individuals can offset the costs of investment in a wind farm against tax. This makes wind farms an attractive investment option, especially for smaller investors, and can lead to a proportion of the capital costs for new wind farm developments being financed by the general public.

• *Financial: Subsidies and low-interest loans are available to wind energy projects*

Many German regions provide financial support to renewable energy schemes, including investment subsidy programmes, to implement their energy policies. Nordrhein-Westphalia, for example, actively supports renewable energy, and in 1997 alone provided some EUR 10 million to financially support environmental energy projects, including some wind energy projects.

A number of German financial institutions provide low-interest loans suitable for renewable energy projects. For example, the European Recovery Programme (ERP) — Environment and Energy Savings offers long-term loans with low interest rates for investments in the use of renewable energy. The programme is administered by the federally owned Deutsche Ausgleichsbank (DtA), Bonn. Loans may amount to 50 % of investment costs, and provide favourable interest rates and loan arrangement. Loans to wind energy projects over the 1990–97 period reached DEM 3.48 billion (EUR 1.78 billion) out of the DEM 4.18 billion (EUR 2.13 billion) disbursed to renewable energy projects.

The Halde Nierchen scheme received no grants but does benefit from a soft loan from the Nordrhein-Westphalia regional government under its Rationelle Energie Nutzung (Rational Use of Energy) programme.

• *Administration: Planning guidance is being developed in some regions to identify areas for wind developments*

Public acceptance is becoming a problem in areas with a number of wind turbines. To help to overcome local opposition to wind power developments, planning guidance is now being developed in some regions to identify areas open to or barred from wind developments. In addition, national land use directives are in preparation to indicate how much renewable energy should be developed in each of the regions, particularly in regions (such as Nordrhein-Westphalia) where there is a high level of wind energy developments.

The planning issues associated with the Nordrhein-Westphalia scheme were complex, largely because the farm is on the border of two regions and so approval from both regions was needed before the developers could go ahead, which took 3–4 years. There are no plans to strengthen coordination within and between the *Länder* of measures to promote renewable energies.

• *Technological development: A strong and expanding German wind energy industry*

Germany's wind energy manufacturing industry is expanding to meet the increasing demand for domestic installations, both through indigenous companies and through joint ventures, especially with Danish companies. This expansion is made possible because manufacturers are confident of a steady future market based around the continuation of the feed-in law in the Renewable Energy Sources Act.

• *Information, education and training: Active involvement of locals in wind energy projects*

There is generally a high level of environmental awareness amongst German citizens and an interest in wind energy as an alternative to other energy sources. Wind energy is of especial interest to farmers, who see it as an opportunity to provide an alternative income stream through land rentals or through electricity sales.

Many wind farms are part-financed by local community subscriptions. In the Halde Nierchen scheme, the developer held public meetings in local communities to raise interest in the scheme. Just over 30 % of the scheme subscribers are resident in the local area.

## 5.6.  Spain — Photovoltaics

*Electricity from photovoltaics (PV) has recently increased dramatically in Spain, placing Spain among the leaders in PV exploitation in Europe.*

| In 1993: | 1.2 GWh |
|---|---|
| In 1999: | 17.0 GWh |
| Increase 1993–99: | 15.8 GWh, 1 330 % |

Spain has high potential for photovoltaic (PV) energy. PV electricity has been expanding rapidly in particular since 1998 when it received increased financial support through national feed-in tariffs. Financial support by regional authorities is also encouraging PV development in some of Spain's 17 autonomous communities.

**Success factors:**

• *Political: Support for renewable energy implementation at both national and regional level*

The national Energy Saving and Efficiency Plan (PAEE), 1991–2000, aimed to increase the overall use of renewables by 1.1 mtoe by the year 2000, including an increase in the contribution of non-hydro renewables in electricity generation from 0.5 % in 1990 to 1.4 % in 2000. This support is endorsed and implemented at the regional level. PAEE provided funds for energy projects at national level until 1999, when responsibility for distributing the funds was transferred to each autonomous region. The Plan de Fomento de las Energías Renovables (2000–10) sets out new targets and aims to double renewable energy to 12 % of gross inland energy consumption by 2010.

Regional encouragement of PV energy has been particularly strong in the Canary Islands, Andalucia and Castilla la Mancha regions.

• *Legislative: Premium-set tariffs combined with an obligation to purchase provide a stable, commercially favourable market for renewable electricity producers*

A series of royal decrees during the 1990s provided support for electricity generation from renewable energy sources, wastes and combined heat and power. The decrees guaranteed the purchase of electricity from renewable sources at a premium fixed price. The 1994 decree determined the fixed tariff for solar electricity at ESP 10.42/kWh (EUR 0.06/kWh). The 1998 decree drastically increased the fixed tariff for solar electricity to ESP 66/kWh (EUR 0.39/kWh) for PV installations smaller than 5 kW, or ESP 36/kWh (EUR 0.22/kWh) for larger installations. The 1998 legislation also provided guaranteed access to the electricity grid, with agreed rates for connection.

• *Financial: State and regional subsidies available*

The PAEE has provided subsidies in the form of capital grants, initially up to ESP 800/Wp (4.8 EURO/Wp) for grid-connected systems and up to ESP 1 600/Wp (EUR 9.6/Wp) for non-grid-connected systems. Since 1996 PAEE has provided subsidies up to ESP 600/Wp (EUR 3.6/Wp) and ESP 1 200/Wp (EUR 7.2/Wp) respectively. Since 1997 only certain regions of Spain have been eligible for grants.

In addition to PAEE the autonomous regions have established support for investment and financing of renewable energy projects. Programmes such as Prosol and Procasol provide capital incentives for the provision of energy to isolated rural communities, hotels and leisure centres in Andalucia and the Canary Islands. Technologies such as PV home systems are eligible for support under these programmes.

For example in Andalucia, subsidies of EUR 11.9/Wp for non-grid-connected systems are available and EUR 8.92/Wp for grid-connected installations.

- *Administration: Local involvement in renewable energy planning combined with collaboration among local, regional and national administrations*

Responsibility for renewable energy sources belongs chiefly to the autonomous communities (the regions). In particular, this allows each region to have authority over the various administrative procedures including planning and environmental impact assessments, to implement renewable energy projects.

Successful implementation of PV projects is mostly met where collaboration at all levels of administration (local, regional and national) is achieved.

## 5.7.  Spain — Wind energy

*Spain is rapidly becoming one of the leaders in wind power exploitation in Europe.*

| In 1993: | 116 GWh |
|---|---|
| In 1999: | 2 744 GWh |
| Increase 1993–99: | 2 628 GWh, 2 266 % |

Spain has high wind energy potential. Wind energy development is expanding considerably, mostly due to financial support through national feed-in tariffs but also because of capital subsidies, in particular at the beginning of a development. Expansion may, however, be held back by lengthy procedures to obtain planning permission.

Navarre was one of the first regions to actively support wind developments, opening the way for other Spanish regions. Spain now has successful wind energy developments in many regions.

El Perdon wind farm is one of the projects developed in Navarre by Energia Hidroelectrica de Navarre (EHN). The first phase, of six 500-kW wind turbines, came into operation in 1994 as an initial demonstration. More turbines were installed in 1995–96, and total installed capacity is now 20 MW. Based on its early successes, EHN is now a leading developer of wind energy projects throughout Navarre and other regions.

**Success factors:**

- *Political: Support for renewable energy implementation at both national and regional level*

The national Energy Saving and Efficiency Plan (PAEE), 1991–2000, aimed to increase the overall use of renewables by 1.1 mtoe by the year 2000, including an increase in the contribution of non-hydro renewables in electricity generation from 0.5 % in 1990 to 1.4 % in 2000. This support is endorsed and implemented at the regional level. PAEE provided funds for energy projects at national level until 1999, when responsibility for distributing the funds was transferred to each autonomous region. Spain has already reached the 2000 renewables target. The Plan de Fomento de las Energías Renovables (2000–10) sets a new target of 12 % share of renewables in gross inland energy consumption by 2010.

The regional government of Navarre developed an energy plan in 1996 with the aim that by 2005 all electricity generated in Navarre will come from renewable energy sources, of which about 50 % will be from wind energy.

- *Legislative: Premium-set tariffs combined with an obligation to purchase provide a stable, commercially favourable market for renewable electricity producers*

The main driving force behind renewable energy comes from a series of royal decrees during the 1990s on support for electricity generation from renewable energy sources, wastes and combined heat and power. The decrees guaranteed the purchase of electricity from renewable sources at a premium fixed price, at 80–90 % of the average electricity tariff from conventional power sources. From 1999, wind electricity producers could receive either the fixed tariff of ESP 10.42/kWh (EUR 0.06/kWh) or the average hourly market price of electricity plus a bonus of ESP 4.79/kWh (EUR 0.03/kWh). The legislation also provides for guaranteed access to the electricity grid, with agreed rates for connection.

- *Financial: State and regional subsidies available*

The PAEE provided subsidies in the form of capital grants, up to 30 % of eligible costs of the project. In addition, each autonomous region has established relevant support for investment and project financing. Because of the success of wind energy in Spain, support available through the PAEE was reduced during the PAEE period (1991–2000), both for the systems and regions eligible for capital subsidies, and for the subsidy provided per project.

- *Technological and industrial development: A thriving wind manufacturing industry has been established in the region*

Decreasing costs of investment and operation, and the greater maturity of the technology, have greatly helped the development of Spain's wind energy industry.

Navarre is a highly industrialised region, and support for new wind energy developments was seen to provide not only environmental benefits but also economic benefits through new employment and general economic development. As a result of the high level of new wind farm developments in Spain, especially Navarre, Gamesa Eolica is now the major manufacturer of turbines in Spain and one of the world market leaders in the sector.

The installation of the first phase of the El Perdon wind farm led to the set-up of three factories in Navarre — one for blades, another for towers and a third for turbine assembly. The wind industry in Navarre now provides employment for over 1 000 people in the region. EHN has expanded its operations to other regions of Spain, installing 211 MW new wind capacity during 1999 (total capacity reaching 418 MW in that year). The aim for 2000 was to install a further 480 MW, and to achieve 1 000 MW new capacity each year between 2001 and 2003.

- *Administration: Local involvement in renewable energy planning*

Responsibility for renewable energy sources belongs chiefly to the autonomous communities (the regions). This allows each region to have authority over the various administrative procedures, including planning and environmental impact assessments, when renewable energy projects are implemented.

The Navarre regional government actively supported the development of procedures to authorise wind farms in the region. It established the public–private company EHN to develop the region's renewable energy resources. EHN's shareholders include the government, the regional electricity supply company, local industry and the regional bank. In June 1996, the government of Navarre approved EHN's Wind Power Development Plan for the region, with a target to install 575 MW by 2010.

- *Information, education and training: Active promotion of the benefits of renewable energy by wind energy developers*

Wind energy developers, including EHN, actively carry out consultations with a very wide range of interested parties before establishing new projects. Consultees vary widely and include municipal councils, conservationists and mountaineering organisations. There are also ongoing campaigns both by the private developers and by the municipalities to provide information to the general public about the benefits of wind energy and the status of the region's energy plan, to ensure continued public awareness and support.

## 5.8. Sweden — Biomass district heating

*Forestry is one of the most important natural resources in Sweden, which has a long history of making use of this resource for fuel.*

| In 1993: | 365.9 ktoe |
| In 1998: | 571.3 ktoe |
| Increase 1993–98: | 205.4 ktoe, 56 % |

District heating systems are widespread in Sweden, with over one third of the total domestic heat market supplied from district heating. There are nearly 200 plants, and biomass is one of the main fuel sources. Biomass as a fuel source for district heating plants has been increasing steadily over the past two decades, particularly to replace electricity for heating. Biomass resources now meet more than 50 % of the fuel supply to district heating networks.

**Success factors:**

• *Political: Support for renewable energy use, especially biomass*

The overall objective of Sweden's energy policy is to secure the long- and short-term energy supply on economically competitive terms, with an emphasis on sustainable development. Sweden has a policy to prevent an increase in carbon dioxide emissions, and it has also made commitments to phase out its nuclear generation capacity.

Long-term support for research and development into new and renewable energy technologies, and a greater use of renewable energy, are two principal means of achieving these aims. Biomass especially plays a vital role. Sweden has a policy objective to replace electric domestic heating with combined heat and power or district heating systems, especially making use of biomass for fuel.

• *Fiscal: Tax system benefits biomass use*

Biomass is exempted from the energy tax, the carbon dioxide tax and the sulphur oxides tax. The increase in biomass district heating has been greatly helped by the introduction of carbon and energy taxes as their application made other options, in particular coal-fired district heating plants, more expensive.

• *Technological development: Active development and promotion of biomass technologies*

Swedish research and development actively supports technological developments in renewable energy. Biomass research, development and demonstration receive total funding of about SEK 400 million (EUR 35 million) per year from the government. Electricity companies and other industries also provide funds. The main areas of support are combustion and conversion technologies, demonstration of pre-competitive technologies, fuel production, harvesting supply programmes and ash recycling.

• *Administrative: Municipalities actively support the establishment of biomass district heating systems*

Development of biomass district heating systems is primarily the responsibility of each municipality. Most domestic district heating systems are owned and operated by municipalities, or by private companies on their behalf. Biomass-fuelled district heating provides economic and environmentally sustainable heating for domestic and industrial use, whilst at the same time providing economic benefits through employment of the local population and a disposal option for sawmill wastes. A number of municipalities have recognised the socio-economic and environmental benefits from biomass district heating and are proactively promoting biomass-fuelled systems.

# 6. Analysis of Member State/technology examples of successful penetration

**Photo:** Volker Quaschning

The previous section, together with further examples provided in Annex 1, present a series of Member State/technology combinations. They examine the influence that potential success factors have had on the implementation of a technology in a Member State. The potential impact that each factor has on successful penetration is assessed below, by drawing together the results from Section 5 and Annex 1. Based on this analysis, the lessons learnt are shown in Section 7. Each potential success factor (political, legislative, fiscal, financial, administrative, technological development, and information, education and training) is discussed in turn below.

## 6.1.   Political
The examples shown in Section 5 and Annex 1, demonstrate that strong political support at national, regional or local level is a consistent component in successful penetration of renewable energy in each of the Member State/technology combinations studied.

**National policies in support of renewable energy**
Each of the examples was implemented in a Member State which demonstrated strong support for the development either of renewable energies in general or of a particular renewable energy. By the end of the 1990s, most governments had implemented energy plans

that supported the development of renewable energy and/or had identified national or regional targets for increased renewable energy use and associated policies and measures to support renewable energy uptake.

The main reasons for supporting the development of renewable energy lie in national energy policies which aim to encourage diversity and security of supply, to reduce imports of fuels and to reduce greenhouse gas emissions (in particular carbon dioxide). Renewable energy can make an important contribution towards achieving these objectives. At the point of generation, wind and solar sources provide energy without any associated carbon emissions, while biomass energy is carbon-neutral, provided that the carbon released is recycled in the form of new biomass growth.

Some Member States started to develop renewable energy support programmes earlier than others, usually for country-specific reasons. For example, security of energy supply is given high priority in Member States which have few indigenous fossil fuel resources and must rely heavily on imports. Austria, for instance, has no indigenous fossil resources and has long recognised that renewable energy can help to reduce its fossil imports. Austria's long-established and active political support towards increasing renewable energy use has contributed towards improving the country's security of supply, reducing its energy imports and improving its balance of payments. The Austrian government recognises that this support also improves domestic employment and stimulates indigenous jobs.

Denmark was also quick to identify the potential role that renewable energy could play in employment and job creation. For more than 15 years, Denmark has provided active support to the expansion of its renewable energy industry, through setting targets backed up by practical support measures. The government saw renewable energy, and wind power in particular, as an opportunity to contribute towards a more sustainable fuel mix for energy production (especially by reducing coal use). It recognised that this support would stimulate the development of an emerging industry, enabling Denmark to become a market leader in wind energy. The Danish government therefore implemented a series of energy action plans during the 1980s and 1990s, each becoming progressively more ambitious in terms of renewable energy use, and in line with the government's overall carbon dioxide emission reduction targets.

Another factor acting in favour of increasing support for renewable energy is the political attitude towards nuclear power, especially where there is a national desire to use less or none of it. Sweden has committed itself to phasing out its nuclear capacity, but any replacement with fossil capacity could conflict with the country's international commitments to limit greenhouse gas emissions. Renewable energy, combined with energy efficiency measures, offers an environmentally acceptable alternative.

During the latter part of the 1990s, more and more Member States implemented specific policies towards renewable energy as part of their national energy plans, and developed action plans and targets for increasing their use of renewable energy. Many of these policies and plans are now being expanded.

**Denmark** implemented a series of energy strategies throughout the 1990s which progressively raised the targets for renewable energy use. Its 1996 strategy, Energy 21, set the target of 1 500 MW of wind turbines by 2005. This target was exceeded in 1999 and a new goal of providing 20 % of electricity consumption from renewable energy resources by 2003 was set. The government's longer term ambition is to generate 50% of the country's electricity requirements from renewable energy sources by the year 2030, a large part of which will come from on and off-shore wind turbine installations.

**Finland's** 1994 national biomass strategy aimed to increase biomass use by 25 % between 1992 and 2005. The more recent Action Plan provides a further stimulus with a target to increase renewable energy use, including biomass, by 50 % between 1995 and 2010.

**Ireland's** 1995 targets for electricity generation from renewable energy were increased in a 1999 White Paper to 500 MW by 2005.

**The Netherlands'** 1997 White Paper set goals for renewable energy use which were subsequently made more challenging in 1999; they were to meet 5 % of gross inland energy consumption from renewable sources by 2010 and 10 % by 2020.

**Spain's** 2000 renewables target has now been expanded to a 12 % share of renewables in gross inland energy consumption by 2010.

### Regional renewable energy policies

The case studies in this report highlight the important contribution made by regional energy policies in encouraging renewable energy. For Member States with a high degree of regional autonomy, such as Austria, Germany and Spain, many regional authorities have brought forward plans that are more supportive of renewable energy than those implemented at national level. For example:

— Upper Austria initiated an energy plan in 1994 that set clear targets for increasing the use of renewable energy in the region. All Austrian regions now have similar policies.
— In Germany, many regions have established targets for increasing their level of renewable energy use. Regional wind energy targets have been established in northern German *Länder* such as Nordrhein-Westphalia, and these have helped give these regions the highest levels of wind energy use in Europe.
— A similar approach was taken in Spain, where regional governments have taken the initiative in stimulating renewable energy in their regions. Navarre is the one of the most advanced regions in its support for renewable energy and especially for wind power. Other regions are now also showing high levels of renewable energy use.

## 6.2. Legislative

### 6.2.1. Power purchase

A guaranteed market for power sales was consistently identified in the case studies as a success factor. During the six year period (1993-1999) covered by this study, two principal legislative options were available for power generators in the various Member State/technology combinations studied - feed-in arrangements and competitive tendering.

### Feed-in laws

One of the most effective support measures for encouraging increased renewables generation is the feed-in law, which provides guaranteed power purchase agreements at fixed prices. The most favourable rates and conditions, and consequently some of the greatest rates of increased penetration, especially of wind energy, are to be found in Germany and Spain. In both countries, utilities are obliged to purchase renewable electricity. The prices paid are guaranteed and at a preferential rate.

**Wind:** The success of the feed-in law is clearly visible in the rapid increase in output from wind in two Member States: output in Germany rose by over 700 % between 1993 and 1999 (from 674 to 5 528 GWh), and in Spain by 2 266 % over the same period (from 116 to 2 744 GWh). The pace of new installations has accelerated quickly in both Member States in recent years, corresponding to the implementation of feed-in legislation. Denmark also achieved a long and sustained growth in wind power installation during the 1990s, again due to developers being able to sell their power at a known and economically favourable rate.

These three Member States dominate the wind power market. The scheme is simple, and provides guaranteed and known power prices, over a number of years. The arrangement removes a large amount of the uncertainty and risk associated with the development of a renewable energy scheme. Other Member States which have offered feed-in arrangements have generally not achieved such high take-up rates. For example, when Italy introduced its pricing system, it was very complex; when the arrangements were simplified more projects were brought forward.

**Biomass:** The use of biomass in power stations (including biomass combined heat and power stations) has also benefited from feed-in laws, particularly in Denmark, Germany and Spain. The Danish purchase obligation system for biomass has been in operation since the early

1990s. However, despite a 189 % increase in generation output between 1993 and 1999, growth was lower than anticipated, especially in later years. This was mainly because the price available for renewable energy did not change for a number of years and producers no longer considered it commercially viable. Reforms of the electricity industry currently being implemented will alter this situation, through changes to feed-in tariffs and rules for electricity from renewable energies. For instance, green certificates are currently in preparation in Denmark. In Spain, interest in using biomass to generate power has increased with the introduction of attractive feed-in tariffs for generators. In Germany, the feed-in law was revised in 2000 to provide more attractive rates for biomass generators.

Attractive rates are important to obtain sufficient levels of interest and investment in new biomass power plants, which need high levels of capital investment.

**Photovoltaics**: Successful photovoltaics (PV) implementation has also benefited from feed-in laws. Germany and Spain are the only countries that showed both a high rate and level of penetration over the 1993–99 period. Both countries provide generous feed-in tariffs for PV electricity. Spain revised its tariff upwards in 1998, and this contributed to Spain's increased generation of PV electricity in 1999. In 2000, Germany also revised its PV feed-in tariff upwards. PV technology is at pre-commercial stage for most applications, and cannot yet compete commercially with other energy sources, even with other renewable energies. PV therefore still requires considerable financial support. Germany and, to a lesser extent, Spain have established, in addition to feed-in tariffs for PV electricity, generous funding support schemes to stimulate the level of uptake of PV (see Section 6.4).

In the long term, and as levels of renewable energy increase, the rapid success of feed-in arrangements is leading to concerns over how the costs of the support should be shared among consumers. Germany may be one of the countries where these considerations will be reached most quickly, because it has very high rates of uptake of renewable energy. Some regions in Germany have seen wind capacity grow to represent more than 5 % of the region's total electricity capacity in less than five years. Since 1998, the German electricity market has been opened to competition, and unregulated access was granted to electricity transmission and distribution systems. Under the terms of the 1991 feed-in law, utilities were obliged to pay the cost of the feed-in tariff, passing the extra charge on to the customer. This resulted in high charges for electricity in areas where wind power schemes were common. After 1998, German regulations limited the utilities' purchase obligations for renewable electricity to 5 % of the total electricity consumption within their supply area. When the local utility reached this limit, the obligation no longer applied. The new 2000 German Renewable Law changed this arrangement: it abolished the 5 % limit and introduced a system according to which the costs of the support to renewable energy are distributed among all grid operators so that the costs are borne equally.

**Competitive tendering**
A competitive tendering mechanism is the main alternative to the feed-in system. It provides a guaranteed market through access to contracts and competitive prices for renewable energy. Competitive tendering has been operated, through largely similar systems, in France, Ireland and the UK. The UK did not meet either of the selection criteria for the successful penetration of renewable electricity technologies. Ireland and France both showed rapid rates of expansion of wind energy between 1993 and 1999, although the fact that both countries started from very low initial penetration levels should be taken into consideration.

Ireland established the Alternative Energy Requirement (AER) as a support mechanism for a range of renewable energy technologies, based on the system in place in the UK since 1990 called the Non-Fossil Fuel Obligation (NFFO [14]). Both the AER and the NFFO are competitive bidding systems whereby developers respond to calls for tender ('tranches') to provide electricity from a range of renewable energy sources. If they are successful, they

---

(14) The NFFO is being replaced by a Renewables Obligation, the draft of which obliges suppliers to provide an increasing proportion of their supplies from renewable sources (3 % by April 2003, rising to 10.4 % by April 2011). Suppliers will need to purchase new renewable energy capacity, and it is expected that this will stimulate demand. Increased demand will also encourage premium prices to be paid for renewable electricity.

obtain a guaranteed power price, at the level of their bidding price, and a long-term (e.g. 15-year) contract for power sales from their renewable energy project. Each type of renewable energy project is grouped with other similar technologies, which ensures that there is competition between applications.

France established the Eole system in 1996, which provided support for wind energy through a competitive bidding mechanism, similar to the ones in Ireland and the UK. Its success has been limited. Most of the wind energy developments have been small installations on the island of Corsica or in French overseas departments. Project developers encountered a range of barriers not directly associated with Eole, which have tended to slow down project implementation. Since June 2001 France has replaced competitive tendering with a system similar to the feed-in law to further promote wind energy.

### 6.2.2. Access to the grid

Straightforward access to the necessary grid infrastructure is also critical to success. Renewable energy electricity generation faces problems of grid access that are different or absent for larger generators. Renewable energy is generally small scale, decentralised, and may be located in rural or remote locations where grid connections are limited or unavailable. In addition, much renewable energy is intermittent ([15]) in nature, especially wind, PV and hydro, and this can attract penalties under some grid access charging tariffs, which favour generators that are able to provide continuous and consistent generation output.

Member States that took the biggest steps to address problems of grid access achieved the greatest levels of renewable electricity penetration during the 1990s, especially for smaller-scale renewable energy projects. Denmark, Germany and Sweden all have policies that oblige utilities to allow straightforward access to the grid for renewable energy producers. They also have transparent and economically fair charging systems for grid access, so that developers know the charges that they are likely to face, in advance.

Two barriers have been identified, in the cases examined in this study, which can limit the ease with which a developer can get a new renewable energy project connected to the grid, and the cost of achieving this.

Firstly, the grid itself may have limitations that make it difficult for a renewable energy project to be connected. This is often the case where the grid does not have the capacity to accept new power load, or where the proposed site for the renewable energy project is remote from a convenient grid access point. This limitation to wind expansion was encountered in the Portuguese wind energy case study, as well as in southern Italy, where large regions have inadequate grids. Rectifying this requires grid extension/strengthening, which may be very costly and could make the proposed project uneconomic if the developer has to bear these costs.

The second barrier occurs when the independent developer is not given ready access to the grid, at a reasonable price. This is the case in France where it has been a contributing factor to wind energy levels being considerably lower than those in Germany and Spain. (This problem has also been encountered in the UK, where grid access charges are not fixed or transparent, and may vary considerably between different regions of the country or between different utility companies.)

Access to electricity grids for renewable energy sources has not been treated equally among Member States. This is one issue that the EU directive on renewable energy in the internal electricity market (see Section 7), addresses. The directive requires Member States to take the necessary measures to guarantee the transmission and distribution of electricity produced from renewable resources and encourages such electricity to be given priority access to the grid. Measures to achieve this include developing transparent and non-discriminatory systems and costs for grid connection. The implementation of these requirements in all Member

---

(15) There is no power generated when the wind does not blow/when it is dark/when there is little or no water.

States should help renewable generation to penetrate much more successfully into the EU's electricity networks.

## 6.3.  Fiscal

Fiscal measures may be implemented in the form of environmental taxes, which penalise the use of fossil fuel (and hence benefit renewable energy use) or as a positive form of tax incentive for environmentally beneficial investment, such as a tax exemption or reduction. Both forms of fiscal support measure are increasingly being used by Member States to encourage renewable energy and other environmentally beneficial activities, or for penalising fossil generation or other environmentally damaging activities.

### 6.3.1.  Environmental taxes

Denmark was one of the first countries to implement an environmental tax. Energy consumers were charged a $CO_2$ tax from 1992, with some of the revenue given to generators of electricity from renewable sources.

Member States, including Austria, Denmark, Finland, Italy, the Netherlands and Sweden, have now implemented various forms of environmental tax. They have introduced energy or environmental taxes as part of their overall energy policy to more accurately reflect the total costs (including costs to the environment) of generating heat or power from fossil or non-fossil sources. The types of environmental taxes implemented vary widely; they may be levied on energy use (both heat and electricity), power or heat generation, $CO_2$, or $SO_2$.

In all cases, renewable energy generation for power and heat benefits through either exemptions or refunds. In Sweden, the introduction of carbon dioxide and energy taxes from which biomass is exempted helped the expansion of biomass district heating and biomass combined heat and power plants, as the taxes made other options, in particular coal-fired district heating and coal-fired combined heat and power plants, more expensive.

In some cases (including Austria and Italy), some or all of the revenues received from energy taxes are recycled to benefit renewable energy, energy efficiency, or other environmental projects.

### 6.3.2.  Tax incentives for investment

Fiscal arrangements can also be used to encourage investment in renewable energy or energy-saving projects.

Tax exemptions or reductions can encourage private individuals and companies to consider investing in renewable energy projects as an attractive financial option. For example, in Germany and Sweden, investment in wind schemes can be offset against tax for individuals, while in Ireland, the Netherlands and Spain companies receive tax relief if they invest in renewable energy projects. In Greece, the installation of solar thermal water-heating systems has been stimulated by tax exemptions for households buying renewable energy appliances such as solar water heaters. In the Netherlands, companies and firms which invest in energy-saving projects (including renewable energy projects) can benefit from claiming accelerated depreciation of investment in equipment for such projects.

### 6.3.3.  Biofuels — benefiting from lowering of excise duty

Biofuels in France benefit from a special fiscal arrangement. In the 1992 mineral oils directive (92/81 article 8.2(d), the European Community allowed Member States to lower excise duty rates for technological purposes (for pilot plants) to develop fuels from renewable energy resources. France, in particular, took advantage of this regulation to lower excise tax on biofuels, making biofuels competitive with higher-taxed fossil fuels. For example, French aid to biodiesel (a type of biofuel) was approximately EUR 120 million in 2000, with similar levels of support in previous years.

The future of this type of support is nevertheless unclear following a recent European Community legal ruling (September 2000) that these reduced rates were applied, in the case of France, for economic and industrial rather than technological purposes, and should, therefore not be granted according to article 8.2(d) of the mineral oils directive. However, article 8.4 of the same directive allows Member States to grant derogations (to lower excise duty rates) for other policy purposes too upon Council approval. France has now applied for such a derogation. This tax exemption is being investigated by the European Commission in order to examine the compatibility of France's application for derogation under article 8.4 with Community state aid rules for environmental protection (OJC 37, 3.2.2001).

## 6.4. Financial

Financial support for the development of renewable energy schemes at the commercial or near-commercial stage is found in almost all the Member State/technology combinations studied. This kind of support is generally through grants or loans towards capital or operational costs. It is clearly an important factor in influencing the successful implementation of renewable energy technologies, both for power and for heat technologies.

### Public sector funding

**Wind**: With the implementation of feed-in arrangements to provide guaranteed premium prices for power purchase, there is progressively less requirement for developers to also receive grants towards their installations. The guaranteed prices available through feed-in support give investors sufficient confidence to invest in the market. In some regions in Germany, however, support through the feed-in law has been enhanced by grants or low-interest loans for wind energy (and other renewable energy) developments. The successful and rapid expansion of Germany's wind energy industry is mainly due to the availability of feed-in support rather than as a result of subsidies. In Spain, because of the success of wind energy, support available through the Energy Savings and Efficiency Plan (PAEE) has been reduced.

The situation in the Member States which have had competitive tendering support mechanisms is different. France has not provided additional funding to new wind energy projects over and above the support provided through the Eole mechanism. The NFFO system in the UK was also the primary source of support. In contrast, Ireland has provided a range of fiscal and financial measures to support wind energy developments (and other renewable energy technologies, including biomass). This additional support may have been a contributory factor towards the more rapid penetration of wind energy in Ireland over the period studied.

For Member States such as Sweden without either a feed-in or a tendering system, subsidies are still the main mechanism for supporting wind energy schemes.

**Biomass**: Almost all the examples of biomass installations — either for power (including combined heat and power) or for district heating — received a significant level of subsidy from public funds. Austria, Denmark, Finland, France, Germany, Sweden (for biomass power only) and Spain all provided a variety of grants towards the costs of biomass installations.

The high costs of developing new district heating grids, or of updating or extending existing ones, can be a limitation to the use of biomass in district heating. For this reason, it is common in Member States such as Austria for biomass district heating schemes to receive considerable levels of financial support towards the heating grid. However, Sweden did not use large subsidies to develop its district heating system. The introduction of carbon and energy taxes, from which biomass is exempted, and considerable research and development support, along with a number of other reasons, helped biomass district heating to expand in Sweden.

**Photovoltaics**: PV is not yet an economically competitive technology, and requires considerable levels of subsidy to be implemented. Only Germany and Spain showed both high rates and levels of penetration over the six years 1993–99. This was due to the high levels of grants available for PV installations, combined with the guaranteed and favourable purchase price provided through feed-in arrangements. For 1999–2003, Germany has committed EUR 560 million towards its 100 000 roofs programme, thus providing an important stimulus to the PV industry. This new programme, supported by the more favourable feed-in tariffs that have become available for PV, have recently increased even further the already rapid rates of installation of PV in Germany. Other Member States are also starting up similar programmes (such as Italy's 10 000 roofs programme). The costs of PV have dropped considerably over the past 10 years, and it is anticipated that further cost reductions will occur as the market expands.

**Solar thermal**: Three Member States (Austria, Germany and Greece) achieved significant increases in exploitation of solar energy for heat. In all three, substantial subsidies are being or have been provided to households or industry to install solar thermal. This support is still required because the payback period for solar thermal installations is long. In Austria and Germany, many of the subsidies for solar thermal installations are provided from regional government funds, and are implemented at the local level. Solar thermal is often not included in national energy policy targets because it is a heat technology and can most successfully be implemented at the local level.

## Structural Funds

Support for renewable energy has also been accessed from non-energy European Community programmes, especially the Structural Funds. These funds provide support for project development, training and other key measures designed to reduce unemployment and stimulate economic activity, and are targeted towards the most disadvantaged regions of the EU. In the cases examined Austria, Ireland, Portugal and Spain all made some use of Structural Funds to support renewable energy developments. The current round of Structural Funds (2000–06) is likely to give support to a much larger number of projects with a renewable energy component.

Accessing the Structural Funds and other similar programmes can only be achieved if there is support for renewable energy at national and regional levels, and a recognition of the important role that renewable energy can play not only as an energy resource but also as a contributor to economic and social cohesion. The case study of the Austrian district heating scheme showed that it received generous grants from national and regional governments, as well as from Structural Funds, in recognition of the potential benefits for local farmers of having an additional market for their biomass.

## Favourable loans

Low-interest loans for the development of renewable energy projects may be provided by banks or other private sector financial institutions operating with ethical investment principles. Some German and Dutch financing institutions in particular are able to offer low-interest loans for environmental projects.

More often, however, favourable loans are provided through public sector funding sources. These may be administered either at the national level or regionally. In Germany, a national loan scheme to support PV installations is administered by regional state banks. In Austria, low-interest loans and long-term credits are available from public banks to support biomass installations, in particular district heating schemes. State financial institutions are often more proactive in the development of renewable energy in their regions. In Germany, preferential loans are provided through funds from regional or local sources. In the Navarre (Spain) wind energy case study, the regional bank is a shareholder in the public–private company established to develop the region's wind energy resources.

| Table 7 | Summary of the principal support mechanisms identified in the case studies for renewable energy power developments (1993–99) | | |
|---|---|---|---|
| **Member State** | **Legislative support** | **Fiscal initiatives** | **Financial support** |
| Austria | Guaranteed prices | Energy taxes on gas and electricity; revenue partly recycled to support renewable energy | Public grants, subsidies and loans |
| Denmark | Purchase obligation + premium guaranteed prices | Energy/carbon dioxide taxes on fossil fuel; revenue partly 'recycled' to support renewable energy | Subsidies provided historically for research, development and demonstration (R,D&D), especially in wind Subsidies (capital grants — biomass) |
| Finland | Transmission costs are fixed; grid access is open to all producers | Energy/carbon dioxide taxes on fossil fuel; revenue partly 'recycled' to support renewable energy | Subsidies on investments and equipment (capital grants) |
| France | Competitive tendering (Eole 2005) | – | Subsidies (capital grants — biomass) |
| Germany | Purchase obligation + premium guaranteed prices | Tax benefits for investing in renewable energy | Subsidies and low-interest loans for all renewable energy projects provided by local banks Direct financial support: PV roofs programme; subsidies for biomass installations |
| Greece | Purchase obligation + premium guaranteed prices | Tax benefits for investing in renewable energy | Subsidies (capital grants — solar thermal) |
| Ireland | Competitive tendering (Alternative Energy Requirement, AER) | Tax benefits for investing in renewable energy | Subsidies (capital grants) |
| Italy | Purchase obligation + premium guaranteed prices | Energy/carbon dioxide taxes on fossil fuel; revenue recycled to support renewable energy | – |
| Netherlands | Purchase obligation | Energy/carbon dioxide taxation favourable towards renewable energy Tax benefits for investing in renewable energy | Government subsidies Obligation for utilities to invest in renewable energy projects |
| Portugal | Purchase obligation + premium guaranteed prices | – | Interest-free loans Support for grid connection |
| Spain | Purchase obligation + premium guaranteed prices | Tax benefits for investing in renewable energy | Capital grants (biomass) |
| Sweden | Purchase obligation | Renewable energy pays lower or no energy tax or nitrous oxide levy Tax benefits for investing in renewable energy | Investment grants |

## 6.5.  Administrative

**The role of the municipality/regional government**
It is increasingly recognised that the successful replication of renewable energy projects can only be achieved on a wide scale if there is active support at the level at which individual projects are brought forward for approval. In most cases this is at local level.

Public acceptance of renewable energy, and an understanding of the benefits that can come from it, is therefore vital if there is to be a sustained flow of projects accepted for development. A strong commitment from regional or local government is one of the ways that this can be achieved.

Most of the examples of projects or technologies studied in this report received a strong level of support at the regional or local level, from the government or local councils. For example:

— Austrian district heating plants are supported by municipalities, including providing assistance for planning issues and by acting as a consumer for the heat supplied.
— In Germany, some municipalities work with the utility to establish solar thermal installations.
— Municipalities support the use of biodiesel in urban areas in France.
— In Sweden, a number of municipalities have recognised the socio-economic and environmental benefits of biomass district heating and are proactively promoting biomass-fuelled district heating plants.

This local or regional support is vital to successful implementation of many of the technologies, especially the smaller ones such as small-scale wind, solar thermal or biomass heating schemes. The municipality or the regional government has responsibility for overcoming many barriers at local level in order to bring forward renewable energy successfully. These actions include:

— implementing regional legislation in support of renewable energy (see Section 6.1);
— providing funding support for local renewable energy projects (see Section 6.4);
— identifying areas in the region where renewable energy developments are acceptable: wind energy planning in particular benefits from clear guidance from local authorities on where wind turbines are or are not permitted — in Germany, for example, some regional authorities identify appropriate sites;
— ensuring that planners receive adequate information to reach balanced decisions about new renewable energy proposals: local resource assessments help to identify the most favourable sites for new developments — in Ireland, for example, a wind energy atlas was developed to support regional planning for wind;
— ensuring that planning and development legislation and regulations do not discriminate against renewable energy: this can be important in areas such as buildings regulations as new PV panels need to be sited in a position to maximise solar energy collection.

Local support for and promotion of the project or the whole technology is important to overcome these barriers. A number of the project examples used in the case studies had encountered some level of local opposition prior to project implementation: opposition to new wind energy developments in Germany and to a new biomass district heating system in Austria. In the UK, planning and implementation of renewable energy during the 1990s was led by the national government, with little opportunity for regional or local initiatives, and as a result there was a strong level of local opposition to renewable energy developments. Local and regional involvement in planning for renewable energy is now being encouraged to try to overcome this opposition.

### Recognising the socio-economic benefits from renewables

The socio-economic benefits of new renewable energy projects are increasingly becoming an important component in decisions to implement new renewable energy strategies. Renewable energy technologies provide local jobs, and keep investment in the local economy, especially for biomass projects which provide an additional benefit in stimulating the local biomass fuel industry.

Austria in particular has recognised the significant benefits that renewable energy technologies can provide to a region's economy. Many Austrian district heating schemes are supported through Structural Funds, because they help to create employment. Farmers and others in biomass-related employment, benefit considerably from biomass-fired heat or power schemes, to which they can provide readily available fuel.

In Spain, the Navarre regional government supported the development of renewable energy (especially wind) because of its benefits to the local economy. The region has gained not only from inward investment and employment in the installation of wind turbines, but the regional government has supported the establishment of Spanish wind turbine manufacturing capacity, in collaboration with Danish turbine manufacturers. The result has been a very rapid increase in Spanish manufacturing capability for wind turbines, both to service the domestic market and increasingly for overseas markets.

## 6.6.  Technological development

Until the 1990s, renewable energy technologies were less technically advanced than conventional energy technologies. Research and technological development has been carried out to address this. Public sector funding programmes for research, development and demonstration stimulate the development of new renewable energy technologies to make them more competitive in the wider energy sector. These programmes support activities to reduce capital and operating costs, improve efficiency and demonstrate long-term reliability.

Over the period examined, the European Community provided support towards the development of renewable energy through a range of energy programmes such as Joule (research and development), Thermie (demonstration) and Altener (a programme specifically targeted at overcoming non-technical barriers and implementation of renewable energy). The Thermie programme in particular provided support for a large number of renewable energy demonstration projects during the 1990s. Examples included wind farms in Ireland, biomass power plant in the UK, PV installations in Italy and the Netherlands, and biomass combined heat and power in Italy.

Current European Community programmes provide support for renewable energy development, particularly through the fifth framework programme for research and development, and the Altener renewable energy programme. In addition, these and other programmes support newer emerging renewable energy technologies such as wave power and offshore wind power.

National energy research, development and demonstration programmes have also played a vital role in the development of renewable energy technologies. Many of the examples studied highlighted how important early and targeted support for research, development and demonstration at the Member State level has been.

The extent to which each Member State is prepared to support technological developments is often reflected in its subsequent indigenous industrial capabilities and expertise as the following examples illustrate:

— The Finnish and Swedish biomass industries are among the most successful in the EU; much of this can be attributed to the comprehensive research and development support provided by the respective national governments, in collaboration with industry.
— Danish wind turbine manufacturing capability was developed with early support from the government during the 1980s, and Denmark now leads the world in its technological capabilities in this field. During the early stages of development, subsidies were high enough to enable the industry to establish itself. Then subsidies were replaced by financial support for generation output.
— Support for biomass technological development in German regions has helped to establish a strong indigenous capability to service the emerging domestic market.
— Germany and the Netherlands have an indigenous PV cell manufacturing capability thanks to government and industry research in this area.

## 6.7.  Education, information and training

Education and information to the general public is a vital component of a successful renewable energy deployment programme.

Some Member States already derive a high proportion of their energy from renewable sources, especially from biomass resources. Austria, Finland, Portugal and Sweden have some of the highest levels of use of biomass for energy in Europe. In such situations, new biomass projects seldom encounter opposition. They are supported because the population is fully aware of the local benefits — particularly through generating jobs for farmers and creating alternative outlets for biomass wastes from sawmill operations.

Cooperative participation in a project is one way to engage the local population in a new renewable energy development. A number of Member States, particularly Denmark, Germany and Sweden, have a long history of cooperative ventures, particularly for farming. Smaller-scale renewable energy projects can be developed and financed cooperatively, and in these Member States this type of arrangement has been successful in bringing forward a large number of new renewable energy projects, especially wind turbines.

However, in many situations there may be less awareness of renewable energy locally. In addition, larger-scale projects are generally not suitable for cooperative involvement. In these cases, developers need to work with the local community to provide information about the nature of the new developments and their potential benefits. A number of the case studies presented, such as wind energy in Spain, show a strong element of informing the local community before the project was fully accepted.

Both national government and local communities play important roles in raising awareness of renewable energy, through information dissemination activities, education and training at school, in the workplace and to the general public. The overall objective is to raise awareness and to ensure that the potential benefits of renewables are effectively communicated.

### The role of energy agencies

A number of the technology/Member State combinations studied were projects that had been brought forward with the support of the local or regional energy agency — for example, in Upper Austria and in Berlin, Germany. Energy agencies play an important part in promoting and raising awareness of renewable energy, and in transferring national objectives to the regions. More and more of these agencies are being established throughout Europe, most often with initial financial support from the European Commission, with the aim of stimulating the expansion of renewable energy locally. They are seen as a means by which national and in particular regional energy plans can be implemented at the local level. They can work with the municipality to raise awareness and work alongside developers and utilities to achieve real and visible results. Smaller-scale renewable energy technologies, especially heat technologies such as solar thermal and biomass district heating but also PV, have benefited greatly from the proactive involvement of energy agencies.

### Environmental awareness

Citizens of most EU Member States are well informed about the benefits of environmental protection, and the important role that renewable energy can play in a country's energy policy. Concern about nuclear power in particular is one of the main reasons why, for example, Austrians, Danes, Germans and Swedes generally welcome renewable energy as an environmentally acceptable alternative. In most countries the general public can perceive and appreciate the positive environmental benefits from renewable energy, and can therefore translate this enthusiasm into strong commitment and support.

# 7. Lessons learned

**Photo:** Volker Quaschning

In light of the analysis carried out in the previous section, this section draws some conclusions on success factors which may contribute to successful penetration of renewables in the Member States.

Overall, no single factor is of overwhelming significance in the successful deployment of renewable energy technologies. Instead, it is the cumulative benefits from a series of supporting measures that determine the extent to which renewable energy is successfully exploited in any one Member State. There are, however, certain essential components which combine to help create an environment in which renewable energy exploitation can succeed.

## 7.1. Energy policies

All Member States now have or are working towards national energy policies that reflect their commitment towards developing renewable energy. The cases examined highlight how vital political support is in order to achieve high levels of penetration of renewable energy technologies.

At the national level, political support ensures that policies are implemented in favour of renewable energy deployment. National policies, combined with targets for increased levels of penetration, endorse support for establishing stable conditions for renewable energy development. This support includes: frameworks for access to energy markets, grid access and

price support mechanisms, all of which are crucial for renewable energy deployment. Funding for research and development programmes is also generally provided at the national level.

In addition, the studies have highlighted how regional energy policies can contribute towards encouraging renewable energy. For Member States with a high degree of regional autonomy, such as Austria, Germany and Spain, many regional authorities have brought forward legislation that is more supportive towards renewable energy than legislation implemented at national level.

There is likely to be even more focus on the importance of integrating renewable energy into regional policies, especially where targeted support programmes such as Structural Funds are key components of regional development. Regional development strategies are increasingly recognising the significance of renewable energy in sustainable development policies. These include not only the role that renewable energy plays in environmental protection, but also its contribution towards economic development, employment and inward investment, particularly for rural or remote areas.

## 7.2.  Legislation

For electricity from renewable sources, the **feed-in law system,** through both the commercially favourable guaranteed feed-in tariffs and the provision of a long–term stable pricing structure, has given a great impetus to renewable energy developments, in particular wind energy. Three Member States (Denmark, Germany and Spain — all countries using this system) contributed 80 % of new wind energy output in the EU-15 over the period 1993–99.

Biomass power generation has also benefited from feed-in tariffs, but not to the same extent as wind. This may be because the tariffs available were less economically attractive than those for wind. Biomass power increased significantly in some Member States (Finland and Sweden) without the support of a feed-in mechanism. Successful biomass development benefits from the availability of feed-in tariffs, but it is also closely linked with other success factors, especially the availability of financial support. In those Member States that adopted the feed-in mechanism, biomass use has expanded most when capital subsidies were provided as well as feed-in tariffs.

The PV sector is not yet able to compete on a commercial basis against other renewable or fossil energy sources and therefore needs to receive subsidies to stimulate its expansion. Successful implementation of PV requires both a feed-in support for power output and a capital subsidy in order to stimulate market expansion. Those Member States (Germany and Spain) that have instigated this combination of support saw significant levels of PV deployment.

The main alternative mechanism for support to renewable energy, the **competitive tendering process,** has not been as successful as a single support mechanism in achieving rapid deployment. Indeed overall levels of renewable energy use in countries where this system has prevailed are significantly lower than levels in countries with the feed-in law system. The UK was the first to develop a competitive tendering system, through its NFFO process, but despite early expansion in renewable energy capacity during the mid-1990s, it did not continue to show such rapid growth, which is why it did not pass any of the criteria for successful penetration for any of the electricity technologies studied ([16]). Similar competitive tendering systems have been in operation in Ireland (the AER) and France (Eole) ([17]). For both of

---

(16) NFFO supported electricity from wind, hydro and a range of biomass sources. PV was not included. The NFFO is being replaced by a Renewables Obligation, by which suppliers are legally obliged to provide an increasing proportion of their supplies from renewable sources.

these countries, the case studies indicated that additional barriers, including grid access (France) and infrastructure (Ireland), may have hindered a more rapid expansion in wind energy deployment.

Long-term financial stability is crucial to attracting investor confidence in new installations. Guaranteed tariffs, through feed-in arrangements, provide this degree of confidence, whereas a competitive tendering system opens up uncertainties. Tendering processes generally have uncertain timescales and tariffs, and developers are also unsure whether they will be successful when they bid through the tendering system.

The newly adopted EU directive ([18]) on renewable energy and the internal electricity market suggests that the feed-in mechanism and other national direct support measures may, if necessary, be replaced by EU-wide support measures within the next decade. These may also include the implementation of green certificates, as well as tax exemptions and other fiscal measures. Some of these are in place or under development in a number of Member States. Lessons learnt from the success of the feed-in approach to date, in particular the importance of a guaranteed income and risk reduction to developers, could be useful to the design of any such replacement measures.

Legislation providing **fair access to the electricity grid** for renewable projects is important to ensure rapid uptake of renewable power technologies. Member States that took the biggest steps to address any problems of grid access are the ones where renewable electricity achieved greatest levels of penetration during the 1990s, especially for smaller-scale renewable energy projects. Problems with grid access rights and with fair access charges can be a barrier to easy connection to the grid and can jeopardise the economics of a project. In a few Member States or in more remote regions, such as Portugal or southern Italy, the grid may require strengthening before it can accept additional power load. This may be very costly or limit developments, particularly the expansion of wind energy into more remote areas.

The issues relating to grid access are an important component of the EU directive on renewable energy in the internal electricity market because of the differences between Member States in terms of how they treat grid access for small power producers. The directive requires Member States to take the necessary measures to guarantee the transmission and distribution of electricity produced from renewable sources and encourages such electricity to be given priority access to the grid.

## 7.3.  Fiscal measures

Fiscal measures act in favour of renewable energy in two ways: as part of wider environmental tax initiatives and/or as fiscal arrangements to encourage investment in renewable energy.

Environmental tax initiatives are inter alia implemented as part of Member States' environmental policies, particularly to reduce greenhouse gas emissions. These taxes recognise and reward the environmental benefits from renewable energy, and are increasingly seen as a way of internalising the environmental costs and benefits of energy systems. The case studies demonstrate that environmental taxation is becoming increasingly common in many Member States (Austria, Denmark, Finland, Italy, the Netherlands, Sweden). Taxes are mainly focused on the carbon dioxide or energy content of the fuel, but also on emissions from pollutants ($SO_2$, $NO_x$). Renewable energy benefits either through exemptions or refunds from the taxes, or by being subsidised from revenue raised from the taxes. In Sweden, the introduction of carbon dioxide and energy taxes, from which biomass is exempted, helped both biomass district heating and biomass combined heat and power plants to expand, as the taxes made other options, in particular coal-fired district heating and coal-fired combined heat and power plants, more expensive.

---

(17) Eole supported electricity from wind only. Since June 2001 Eole has been replaced by a system to support wind similar to the feed-in law system.

(18) Directive of the European Parliament and of the Council on the promotion of electricity produced from renewable energy sources in the internal electricity market (2001/77/EC).

Many Member States also allow tax exemptions/reductions or tax incentives for individuals or companies to invest in renewable energy projects. Again, this approach rewards investment in clean technologies, and can also encourage a greater level of awareness of and investment in renewable energy by business or the public.

## 7.4.  Financial support

Financial support to renewable energy schemes is found in almost all the Member State/ technology combinations studied, and is clearly an important factor in influencing the successful implementation of renewable energy technologies, for both power and heat. Nevertheless, in the case of wind energy, when feed-in arrangements are implemented there is progressively less requirement for developers to receive grants towards their installations. The commercially favourable guaranteed prices available through feed-in support give investors sufficient confidence to invest in the market without needing further financial support.

Some national, regional and municipal governments have recognised the potential economic and social benefits of providing support to renewable energy projects, as well as the related environmental benefits. As a result, the EU Structural Funds and other, non-energy, EU funds, are being accessed. For example, biomass district heating schemes have expanded considerably in a number of regions in Austria by receiving targeted financial support through, inter alia, the Structural Funds.

A number of the projects identified in the case studies were supported through banking institutions. The perceived risks of investing in a renewable energy project are reduced if successful examples of the technology already exist. Thus in Member States where renewable energy technologies are well established, there are now financial institutions willing to provide favourable loans towards renewable energy and other environmental projects, because they are confident of their financial viability. The role of the regional government is also important at the early stages of technology development. In these instances, the regional government can create the necessary framework conditions to encourage initial investment by banks in the technology.

## 7.5.  Administration

Most of the successful Member State/technology combinations are backed by strong support for renewable energy developments not only at national level but at the level of local or regional administration. Renewable energy can be deployed successfully only if there is active support and public acceptance for renewable energy at the level at which individual projects are brought forward for approval, in most cases at the local or regional level. A strong commitment from regional or local government is one of the ways that this can be achieved.

Renewable energy schemes can help to stimulate local investment, employment and social cohesion, especially in rural or remote regions. Local authorities which recognise these benefits are generally the ones which provide the most active support towards renewable energy implementation.

## 7.6.  Technological development

A number of Member States have provided strong and consistent support towards technological development of renewables, often targeted towards a specific technology which is most appropriate for their national circumstances (e.g. Denmark with wind, Finland with biomass). They are now benefiting from this investment in the form of strong and expanding domestic capabilities. For example, Danish companies are world leaders in wind turbine manufacture, providing direct or indirect employment for up to 12 000 people in Denmark, with over 70 % of domestic manufacturing for export (Danish Wind Turbine Manufacturers Association).

There is still a need for further targeted funding for research and development. The European Commission in particular plays an important role through its energy support programmes, which assist research, development and demonstration of new technologies, working with national research programmes.

## 7.7.  Education, information and training

In some Member States citizens have a high level of environmental awareness and a desire to reduce dependence on fossil or nuclear energy sources, and understand the role that renewable energy can play. As the case studies emphasise, they are sometimes also well aware of a specific renewable energy, because it is already used extensively (e.g biomass). However, in some other Member States education and information campaigns are essential to generate interest and informed debate amongst the general public.

The success of a renewable energy project, and its subsequent replication, ultimately resides with its public acceptability at local level. At the local level, emphasising non-energy benefits is an important component in the acceptability of renewable energy, especially its role in providing an income stream and local jobs. Cooperative participation in a project can be a successful way of involving the local population in a new renewable energy development. Cooperative financing of renewable energy projects is common in a number of Member States, particularly Denmark, Germany and Sweden. The role of the developer could also be important to the public acceptance of a new renewable project and requires that developers work with the local community to provide information about the nature of the new developments and their potential benefits. In Spain, the example of wind energy included a strong element of information dissemination to the local community before the project was fully accepted.

**Energy agencies**
The role of regional energy agencies in implementing national and regional policies at the local level is also important. Energy agencies can work with municipalities, utilities, developers and the local population to stimulate interest in implementing renewable energy technologies. A number of successful projects have been brought forward with the support of energy agencies, especially smaller-scale projects such as solar water-heating systems targeted at individual households, district heating systems and PV projects in new buildings. The European Commission has recognised the significance of energy agencies and financially supports their establishment. There are now agencies located in cities, regions and islands that are successfully implementing renewable energy and energy efficiency projects.

# References and bibliography

AGORES. A global overview of renewable energy sources: Information about renewable energy, including EU policy and national strategies and database of legislative support in EU Member States (Eneriure). Available at http://www.agores.org

Biomass Congress 2000. Vision of the OPETs concerning European biomass promotion policies. Co-ordinated by Sodean, Spain. Available at www.sodean.es

ECOTEC, 1999. *The impact of renewables on employment and economic growth.* ECOTEC Research and Consulting Ltd. Altener contract 4.1030/E/97-009.

Energie Cites. *Renewable energy in the European cities,* Altener II City-RES Project, Good practice case studies. Available at www.energie-cites.org

EurObserv'ER Eurobarometers. Available at www.systemes-solaires.com/

European Commission, 1997a. *Energy for the future: Renewable sources of energy,* White Paper for a Community Strategy and Action Plan COM(97)599, 26.11.97.

European Commission, 1997b. *EU policy on renewable energy: opportunities in a growing market.* Conference background paper.

European Commission, 1999a. European Union Energy Outlook 2020, *Energy in Europe,* special issue November 1999.

European Commission 1999b. *Energy for the future: Renewable sources of energy,* Campaign for take-off. Commission Services Paper, DG XVII, doc. SEC(99) 504, 9.4.99.

Eurostat. Statistical Office of the European Union. *Renewable energy sources statistics in the European Union,* Data 1989-1998, 2001 Edition.

EVA, 1998. Feed in tariffs and regulation concerning renewable energy electricity generation in European countries, Austria. Available at www.eva.ac.at/

Hassan, Garrad, and Lloyd, Germanischer, 1995. *Study of offshore wind energy in the EC,* EC JOULE contract.

OECD/IEA, 1998. *Renewable energy policy in IEA countries, Volume II Country reports.*

# Annex 1: Other interesting Member State/technology examples of successful penetration

## Denmark — Biomass power

*Denmark has a good record of utilising its biomass resources, especially for heating purposes. More recently, biomass has increasingly been used to generate power as well.*

| | |
|---|---|
| In 1993: | 167.0 GWh |
| In 1999: | 483.0 GWh |
| Increase 1993–99: | 316.0 GWh, 189 % |

Community heating systems are common in Denmark, and biomass is often used as a fuel for them. These systems may be heat only, but increasingly they are being converted or developed to operate as combined heat and power (CHP) systems, with power exported to the grid.

The Masnedø CHP plant produces heat for Vordingborg's district heating network and electricity for eastern Denmark's grid. The plant has a capacity of 8.3 MW of electricity and 20.8 MW of heat. It went into operation in 1996 as one of 10 small-scale CHP plants in eastern Denmark which use indigenous fuels: straw, wood (wood chips), waste and natural gas. The Masnedø CHP plant consumes 62,000 tonnes straw annually, together with natural gas. Wood chips can replace part of the straw if needed.

**Success factors:**

- *Political: Denmark has a national energy plan with targets for increased use of biomass*

The Danish government has implemented a series of energy action plans, the most recent being the current plan Energy 21, launched in 1996. This plan is consistent with the overall national objective of a 20 % reduction in $CO_2$ emissions from 1988 levels by 2005. The plan proposes to increase the use of biomass for energy from 1 250 ktoe per year in 1995 to 2 000 ktoe per year by 2005. One of the main initiatives is to increase the use of biomass in power production, partly through the conversion of biomass in heating-only plants to biomass in CHP, and partly by the development of biomass CHP for smaller district heating plants.

- *Legislative: Premium-set tariffs combined with an obligation to purchase provide a stable market to renewable electricity producers*

Electricity from biomass is sold to the local utility under a feed-in law, in operation since 1992, which guarantees a commercially favourable fixed tariff and market for electricity generated from renewable energy sources.

Utilities are obliged to purchase and distribute electricity produced from biomass sources. The producer pays the cost of grid connection, whilst the transmission company pays for the necessary reinforcement and extension of the network. Prices available vary depending on the time of day and whether the producer is a private company or a utility, and are enhanced through 'reimbursements' of $CO_2$ tax (see below) and government subsidy. In 1998, the average price, including the feed-in tariff, the $CO_2$ tax reimbursement and the subsidy, available for private company biomass energy projects was DKK 0.54/kWh (EUR 0.073/kWh). Local power distribution companies are also obliged to provide grid connection for proposed renewable energy projects.

The increase in CHP installations, although significant, has not been as rapid as was expected, however, because prices have not been increased since the early 1990s. A number of the older plants have experienced economic difficulties, and fewer new plants are coming on line than had been anticipated because the prices are no longer attractive. Reforms of the electricity industry currently being implemented will alter this situation, through changes to feed-in tariffs and rules for electricity from renewable energy sources.

Under a 1993 Biomass Agreement, utilities are also required to use an increasing proportion of straw and wood chips as fuel in their power installations. The Biomass Agreement was amended in 1997 to allow utilities greater flexibility in the choice of biomass procured.

• *Fiscal: Biomass power benefits from favourable taxation*

One component of the beneficial tariffs available for renewable energy projects, including biomass, is a subsidy through the $CO_2$ tax. This is worth DKK 0.1/kWh (EUR 0.013/kWh). Biomass power is, however, subject to $SO_2$ tax (with some compensatory refunds, depending on the size and type of the project).

• *Financial: A wide range of subsidies and grants are available for the biomass sector*

A range of economic incentives is available for the development of biomass projects. These include a fund established in 1992 to support the conversion of district heating plants to biomass-fired CHP plants, which provide 10–25 % of the costs of conversion. Other support is given as subsidies towards the construction of renewable energy projects (up to 30 %), towards demonstration projects and to support the dissemination of commercially available technologies.

## Denmark — Wind energy

*Denmark has led the way in exploiting wind energy over the past 20 years, and until the mid-1990s had the greatest use of wind energy in Europe.*

| In 1993: | 1 034.2 GWh |
|---|---|
| In 1999: | 3 029.0 GWh |
| Increase 1993–99: | 1 994.8 GWh, 193 % |

Many of the wind power installations in Denmark have been developed through the cooperative system, or by local farmers or groups of individuals wishing to develop their own renewable energy power systems. Many of the original turbines are now being updated or replaced with larger machines.

One example is the Avedøre wind energy cooperative, which was established in 1993. Twelve turbines, each of 300-kW capacity, were developed, six of which are owned by the local utility and the other six by a cooperative of citizens from Copenhagen. The wind energy cooperative organised the project and offered Copenhagen citizens the opportunity to become shareholders. The scheme has been successful and has been replicated across the country.

**Success factors:**

• *Political: Long-established political support towards renewable energy development*

Danish energy policy seeks to replace electricity produced from coal with that from CHP, natural gas and renewable energy. Throughout the 1990s, a series of energy strategies has progressively raised the targets for renewable energy use. The most recent (1996) strategy, Energy 21, set a target of 1 500 MW of wind turbines by 2005. This target was exceeded in 1999. The government now intends to provide 50 % of Danish electricity consumption from renewables by 2030. A large part of this will come from on and offshore wind power.

• *Legislative: Premium-set tariffs combined with an obligation to purchase provide a stable, commercially favourable market to renewable electricity producers*

The feed-in law has been an important influence in the development of wind energy projects in Denmark. Electricity from wind farms is sold to the local utility under a feed-in law, in operation since 1992, which guarantees a commercially favourable fixed tariff and market for electricity generated from renewable energy sources. The local utility is obliged to take all the output from the plant. Prices available vary depending on whether the producer is a private company or a utility, and are enhanced by 'reimbursements' of $CO_2$ tax (see below) and government subsidy. In 1998, private and decentralised electricity producers received an incentive of DKK 0.17/kWh (EUR 0.022/kWh), and all renewable producers (including utility companies) received an additional DKK 0.10/kWh (EUR 0.013/kWh) as an internalisation of the external costs of fossil fuels ($CO_2$ tax). A new pricing system was introduced in the year 2000 with the Danish electricity reform and after a transition period of 5 to 10 years, all wind power producers will receive only the market price and green certificates.

• *Financial: Subsidies were available to establish a strong domestic wind energy market*

The Danish government provided subsidies of up to 30 % towards wind turbine installation costs, although these were discontinued in 1989. This early support helped to establish a strong home market and associated indigenous industry.

Grants for replacement of old wind turbines are also available.

• *Fiscal: Taxation is favourable towards renewable energy*

A $CO_2$ tax is levied on electricity production from fossil sources. Renewable energy receives compensation from this, in order to internalise the external costs of fossil fuels (see above).

For cooperative operations, no income tax is payable on dividends up to DKK 3 000 (EUR 400).

- *Administration: Positive municipal support combined with active involvement of local utility*

The long history of using wind as an energy resource has created a favourable environment within the administration for further wind developments. For example, the Avedøre wind energy cooperative project benefited greatly from the positive attitude of the local municipality and the joint development of the project with the local utility. These factors contributed to the simplification of many planning, infrastructure and quality issues.

Since 1996, in order to enlarge the potential of suitable sites, municipalities have been required to submit proposals for wind turbine capacity and to thus make wind turbine development a regular feature of their planning.

- *Technological development: Early government support helped establish a strong indigenous wind energy manufacturing industry*

During the late 1980s the Danish government provided financial support to development projects in the emerging Danish wind energy industry which is now the strongest in Europe, with successful domestic and worldwide export sales.

Domestic wind developments (such as the Avedøre scheme) benefit from this success by having ready access to high-quality Danish turbines, components, sales and service.

Some subsidies are still available from research funds, including the programme for new renewable energy technologies, which supports non-commercially sustainable renewable energy technologies, including wind, and the energy research programme, which supports the implementation of Danish energy policy.

- *Information, education and training: High level of public awareness of environmental issues combined with traditional Danish approach of forming cooperatives*

The concept of cooperatives is well established in Denmark: dairy and other farming cooperatives have been operating for over a century. It has therefore been relatively straightforward to transfer the cooperative concept to the newly emerging wind industry. The Danish Wind Turbine Manufacturers Association has been actively supporting uptake of wind energy for nearly 20 years and offers a high level of support and education to prospective developers, especially farmers and other individuals wishing to develop wind energy schemes.

In addition, there is a high level of environmental awareness among Danish citizens, who see renewable energy as a safe and clean energy supply option. This supportive attitude has helped to raise interest in the environmental benefits offered by renewable energy sources.

### Finland — Biomass power

*Finland is one of the leading European countries in the use of renewable energy, in particular through its extensive biomass resources.*

| | |
|---|---|
| In 1993: | 5 644.0 GWh |
| In 1999: | 8 341.0 GWh |
| Increase 1993–99: | 2 697.0 GWh , 47 % |

Combined heat and power (CHP) is used extensively in Finland to provide both heat and electricity for domestic and industrial purposes. CHP plants make use of a range of biomass resources, burning wood waste and agricultural biomass, as well as peat and black liquor. Much of the biomass resource is derived from Finland's pulp and paper industries, which account for about half the country's industrial energy consumption. There is also a well-developed biomass supply industry. District heating systems are also common, with about half the population connected to a district heating network. These are increasingly using biomass as fuel, as well as coal and natural gas.

The Forssa biomass plant is the first CHP district heating plant in Finland fuelled entirely by wood. The plant's boiler is specially adapted for the use of solid biofuels and other biomass fuels. The plant can generate up to 66 MW heat, and started operating in September 1996. The plant produces all the heat and one third of the electrical power required by the city of Forssa for almost the whole year.

**Success factors:**

- *Political: Finland has an energy strategy that supports the development of renewable energy, especially biomass*

Finland's national biomass strategy was launched in 1994 and aimed to increase biomass use by 25 % (1.5 mtoe) by 2005 from the 1992 level. The Finnish energy strategy is to continue to expand the use of wood and other renewable energy resources, both to help meet Finland's Kyoto targets and to contribute towards the security of Finnish energy supply. An action plan for renewable energy sources was launched in 1999 and has the objective of increasing the use of renewable energy, including biomass, by 50 % from the 1995 level, when renewable energy contributed over 20 % of total primary energy demand, by 2010.

Energy planning is actively supported at regional and local levels, with regional strategies updated regularly.

- *Legislative: Transmission costs are fixed, and grid access is open to all producers*

Liberalisation of the electricity market in Finland started in 1995, and included opening grid access to all producers and consumers. The cost of transmission is fixed by law, ensuring transparency and predictability to producers.

- *Fiscal: Economic support measures act in favour of renewables compared with fossil fuels*

Fiscal measures and research and development (R&D) are two central policy approaches used to support market deployment and commercialisation of renewable energy in Finland.

In 1990 Finland introduced a $CO_2$ tax on fossil fuels, which was replaced by a combined $CO_2$ and energy tax in 1994, based on the carbon and energy content of the fuel. Renewable energy was exempted from the tax.

In 1997 this tax was replaced by taxation on electricity at the distribution level, with a refund granted to electricity from renewable sources (EUR 0.042/kWh for biomass). The heat production tax was retained.

Small-scale (less than 1 MW) biomass plants also benefit from a reduction of value-added taxes payable on the plant.

- *Financial: Subsidies are available on investments and equipment relating to renewable energy production and use*

Public funding is important to support the building of new renewable energy plants and is made available through national, regional or local subsidies. National government subsidies of up to 30 % of the investment costs are available to renewable energy technologies, including biomass.

- *Technological development: Extensive research and development capabilities have formed the basis for the establishment of a strong Finnish biomass technology industry*

The Finnish government provides a high level of subsidy and support to research and development to develop an indigenous biomass technology industry. This has helped to create a strong domestic market and a thriving export industry, in particular in combustion technology, boilers and emissions control.

- *Information, education and training: Active promotion of biomass energy through associations, agencies, etc.*

This includes the establishment of regional energy management agencies, working closely with national energy information centres. Finnish associations produce information material and innovative information dissemination schemes, especially in biomass (e.g. the Finnish Bioenergy Association, the Wood Energy Association). These associations also work closely with organisations in other Member States to make Finnish capabilities in biomass widely known. For example, Finnish biomass organisations have formed close links with similar organisations in Wales in the UK to transfer their know-how to encourage uptake of biomass use there.

### France — Biofuels (biodiesel)

*France is the largest producer of biofuels in Europe, accounting for 40 % of the total European production.*

| In 1993: | 29.2 ktoe |
|---|---|
| In 1999: | 279.3 ktoe |
| Increase 1993–99: | 250.1 ktoe, 857 % |

France is one of the few countries in the world to give a relatively high priority to the development of biofuels, mainly to support the agricultural sector and for research purposes.

For example, there are four plants producing biodiesel, a biofuel, in France. Oil companies can use this biodiesel as a substitute for normal fossil diesel, up to a level of 5 % for use in private cars. In professional fleets (company cars, buses, etc.) it can replace up to 30 %.

**Success factors:**

- *Political: The French government supports a biofuels production programme*

The biofuel production programme is a financial scheme, operated at the national level, to develop investments for biofuel production.

- *Fiscal: Biofuels benefit from advantageous fiscal measures*

In France, biofuels receive exemption from excise tax on petroleum products at the rate of FRF 2.30/litre (EUR 0.35/litre) of biodiesel and FRF 3.29/litre (EUR 0.50/litre) of ethanol in 2000. French fiscal aid to biodiesel, for example, was approximately EUR 120 million per year (FRF 0.8 billion), supporting 337 000 tonnes of oil equivalent. The excise tax exemption means that biofuels can compete cost effectively with fossil fuels.

- *Technological development: French companies are world leaders in biodiesel production.*

The European leader for production and marketing of biodiesel is the French company Diester Industrie, with an annual turnover of EUR 200 million in 1997–98.

- *Information, education, training: Biofuels are actively supported by local communities*

There is an information network among the various communities with an interest in biofuels: the oil companies (TOTALFINA, ELF) and vehicle manufacturers (PSA, Peugeot Citroen, Renault), professional and trade associations (Sofiproteol, NOVAOL), and national non-profit bodies (ADEME, the French Institute of Petroleum — IFT).

Over 30 local communities are working together as the association 'Club des villes diester' to promote use of biodiesel.

# France — Biomass district heating

*The biomass wood fuel market in France is one of the more successful examples of exploitation of renewable energy sources in the country over the past decade, with biomass used mostly for heating apartment blocks.*

| | |
|---|---|
| In 1993: | 7.9 ktoe |
| In 1998: | 16.9 ktoe |
| Increase 1993–98: | 9.0 ktoe, 113 % |

A number of examples of biomass heating applications can be seen in various regions around France:

- Dole, in the Jura Mountains in eastern France, has a 3.2-MW biomass-fired boiler delivering hot water and heating to 1 800 dwellings and various larger public and private buildings. This supplies more than one third of the energy required by the area, and uses 12 000 tonnes of wood residues annually.
- In Normandy, a 2-MW wood-fired boiler plant supplies heating to 470 houses, a college, a school and a sports centre. The project involved the construction of a heating network, which was developed by a heating company.
- In Bourgogne, a district heating system due for renovation was refurbished with an 8-MW wood-fuelled boiler. This provides heat for up to 3 500 homes, and also provides a market for waste wood from local sawmills.

**Success factors:**

- *Political: Support for biomass energy through the biomass wood and local development plan*

This plan was implemented between 1995 and 1998 by the French state agency for the environment and energy conservation, ADEME. It covers both the development of the biomass wood fuel supply sector and the installation of new automated-feed, wood-fired boilers. The installation of 188 boiler houses in apartments and in the industrial and tertiary sector had been achieved by 1999.

- *Financial: Support provided towards the development of projects*

Support for the development of district heating plants is available in the form of investment subsidies from the local region, while an additional subsidy may be available through EU or national funds.

- In the Dole system, about one third of the funding required was provided from ADEME with further contributions from the regional authority and other regional economic development funds.
- In Bourgogne, support was received from ADEME, the regional council and the EU.
- In Normandy, the heating company received financial support from the regional authority together with an agreed contract to operate the plant over a 24-year period.

### France — Wind energy

*Wind energy use in France expanded 10-fold between 1993 and 1999, but the level of penetration remained low.*

| | |
|---|---|
| In 1993: | 3.5 GWh |
| In 1999: | 36.0 GWh |
| Increase 1993–99: | 32.5 GWh, 928 % |

Most French wind energy developments are small installations in isolated or remote regions, particularly islands such as Corsica, or in French overseas departments.

A smaller number of wind installations have been established on the mainland. One example is the series of wind farms that has been developed around Dunkirk over the past 10 years. A first wind turbine, part-funded by the European Community, was commissioned in 1991, and this was followed in 1996 by a wind farm with nine turbines totalling 2.7 MW. Most recently, in 1999, near Dunkirk in Wideheim, a new development was started which comprises six 750-kW turbines, totalling 4.5 MW.

In France, wind energy was supported during the period examined through the competitive tendering system Eole, established in 1996. The process was implemented in stages: the first stage to contract 15 MW, the next to contract 35 MW and finally to achieve a total capacity of 250–500 MW by 2005. Eole was a similar system to the Non-Fossil Fuel Obligation used in England and Wales: requests for proposals were invited from developers for a certain amount of capacity and successful tenders received a guaranteed market for 15 years at the bidding price for their wind-generated electricity. Tariffs available for the first stage were FRF 0.38/kWh (EUR 0.058/kWh), which fell by about 10 % in further bidding stages.

The progress of project implementation was, however, slow. Despite acceptance under the Eole system, developers still encountered barriers to project implementation (see below).

Until recently there have been few fiscal or financial incentives in support of wind energy developments outside the Eole system. The earliest wind turbine at Dunkirk received EU funding support (50 %), but was developed prior to the Eole system. The Regional Council of Nord-pas-de-Calais contributed a 53 % share in the subsequent wind farm constructed in 1996, while additional financing was received from ADEME (the French national energy agency), Eléctricité de France (EdF) and EU Structural Funds.

Wind energy developments in France have been hampered by the difficulties encountered when attempting to obtain connection to the grid. There are no guaranteed access rights for small generators.

This case study illustrates how the lack of a guaranteed market through access to the grid and the limited availability of financial support can restrict the more widespread uptake of renewable energy technologies. The success of Dunkirk and Wideheim is mainly attributable to the foresight and determination of the developers and the municipalities to achieve successful wind farms. For example, the municipality of Dunkirk worked with the developer to stimulate the establishment of the first wind turbine, with the mayor giving his full support to the project. For the second wind project, the municipality of Dunkirk again played a vital role, promoting the project, selecting and making available the site, participating in its financing and encouraging the Regional Council also to participate.

One result of the first two wind power developments in Dunkirk is that the most recent wind farm (1999) in Wideheim has French-constructed innovative turbines.

Since June 2001 France has replaced the Eole competitive tendering system with a feed-in tariff system. Wind energy producers receive on average FRF 0.46/kWh (EUR 0.070/kWh) depending on real wind speed on the site.

## Germany — Biomass power

*Germany has seen a steady increase in the development of wood-fuelled combined heat and power plants, especially in the more forested southern regions of the country such as Bavaria.*

| | |
|---|---|
| In 1993: | 419.0 GWh |
| In 1999: | 677.0 GWh |
| Increase 1993–99: | 258.0 GWh, 61 % |

After the success of the Electricity Feed-in Law in increasing renewable electricity from wind, renewable energy promotion is now shifting towards the direct use of biomass and the use of biomass for power and for heat production.

Bavaria is the largest of the German *Länder*, with about 35 % of its surface area forested, the highest density in Germany. Wood is used for heat and increasingly for power production. The use of wood for heating purposes is common in the region: almost one third of Bavarian homes use wood as an energy source. There are nearly 100 biomass-fuelled district heating systems installed, including 16 combined heat and power (CHP) plants. These plants consume both wood and straw. Bavarian examples of CHP plants include the following:

- Since 1995, a biomass CHP plant has been operating near Sulzbach-Rosenberg. The plant generates 4.2 MW electricity and 16.8 MW heat. Part of the steam produced is used in the nearby steel works, and more steam is distributed for use in the city's district heating network, supplying nearly 1 000 homes. Fuel comes from a farmers' association, which provides a variety of biomass fuels, including straw, wood chips and wood residues.
- A CHP plant generating 15 MW electricity and 75 MW heat provides steam and electricity to a chipboard manufacturing factory in Neumarkt. The plant burns waste wood (sawdust and chippings) and forestry wastes.
- A biomass-fired CHP plant generating 12 MW electricity and 35 MW heat has been established in Altenstadt; it supplies power to the grid and heat to the neighbouring district heating network.

**Success factors:**

- *Political: National and in particular regional government support towards increasing use of biomass*

German energy policy is closely linked with national policies to support climate protection. Renewable energy plays an important part in this policy, and the government has actively supported financial provision towards renewables, both at national and regional level. Most German regions also have energy policies, targets and support mechanisms designed to encourage the development of renewable energy. For example, the Bavarian government aims to increase the share of renewable energy use to 13 %, up from the present level of 9 % of gross inland energy consumption. Of this, the biomass contribution is expected to provide between 3 and 5 %.

- *Legislative: Premium-set tariffs combined with an obligation to purchase provide a stable market for renewable electricity producers*

Through the Electricity Feed-In Law there is a guaranteed market and fixed price for the electricity produced from renewable energy sources, including biomass. More recently, biomass power has been one of the main beneficiaries of the revised feed-in law, which was implemented in April 2000 and provides more economically attractive tariffs for biomass power.

- *Financial: Some grants available for biomass installations*

Grants for investment in biomass power are limited and include capital subsidies from the federal government and low-interest loans from a public bank, the Deutche Ausgleichbank.

- *Technological development: Strong indigenous industry capabilities in biomass*

Research and development (R&D) support has resulted in the establishment of a strong indigenous capability that benefits from a vibrant domestic market.

Since the 1970s, a considerable amount of R&D has been carried out into biomass harvesting, fuel processing and combustion equipment. R&D support has gone to technological developments in wood pelleting systems and new small-scale biomass gasification or CHP biomass utilisation systems. The use of alternative biomass fuels such as straw and other agricultural residues has also been encouraged.

There is a requirement to guarantee good performance levels and expertise to obtain state investment subsidies; this ensures that the overall quality of installations is high.

- *Information, education, training: Promotion of opportunities from renewable energy use through specialist municipal advice centres*

Specialist municipal advice centres provide information on the use of renewable energies, such as biomass. In Bavaria the regional centre for research into agricultural engineering provides an information resource centre to technical and general public to promote the opportunities from biomass use.

# Greece — Solar thermal

*Greece has the highest per capita use of solar thermal technologies in the EU.*

| | |
|---|---|
| In 1993: | 92.5 ktoe |
| In 1999: | 124.4 ktoe |
| Increase 1993–99: | 31.9 ktoe, 34 % |

Solar collectors have been widely installed throughout Greece since the 1980s. There are a number of factors that make solar thermal technologies attractive for providing an increasing proportion of Greece's energy needs, including the country's favourable climate. There is no district heating tradition in Greece, so typical water-heating systems in households are electricity based, and electricity prices are relatively high compared to the Greek household disposable income.

The island of Crete is one area of Greece that has carried out extensive studies and activities to develop its indigenous renewable energy resources. There has been a high take-up of solar thermal collectors among the population, both for domestic dwellings and in the hotel and tourism sector. To date, about 20 % of Crete households have solar collectors. These are mostly produced and installed by local companies, and supply cost-effective and reliable hot water.

**Success factors:**

• *Political: Support policies for renewable energy use*

Various policies and legislation to support uptake of renewable energy, including solar thermal technologies, have been implemented since 1990. The Operational Programme for Energy, which ran from 1994 to 1999, provided a total of EUR 140 million of public, EU and private funds for renewable energy development.

Crete benefited from this programme through targeted support for renewable energy technology development. Crete adopted an energy policy in 1994 that placed a high importance on the use of renewable energy, and the implementation plan was finalised by the regional energy agency in 1999.

Greece's new Operational Programme for Energy, which began in 2000, supports tax exemptions, loans and third-party financing for renewable energy and energy efficiency in the building sector.

• *Financial: Grants and loans available*

In the early stages of their development, in the late 1970s and early 1980s, the use of solar water-heating systems was stimulated through subsidies. From 1990, Law 1892 provided subsidies up to 45 %–55 % (depending on location) for the promotion of various investments including in renewable energies. Hotels which invested in solar water heaters benefited considerably from this law.

• *Fiscal: Tax exemptions to individuals for buying renewable appliances*

Since 1995 Law 2364 has provided tax exemptions to households buying renewable appliances, such as solar water heaters; 75 % of the purchase value of a renewable appliance is deducted from a person's taxable income.

• *Technological development: Strong domestic manufacturing industry*

The financial support given in the late 1970s and early 1980s to promote the use of solar water heaters created an opportunity, and a market, for the emergence and establishment of a local industry which developed over the 1980s and reached maturity in the early 1990s. Local industry was an important driving force in solar thermal expansion in Greece.

## Ireland — Wind energy

*Ireland's exploitation of its wind resource has seen a 12-fold increase in wind energy generation in the six years since 1993 but from a low starting point.*

| In 1993: | 15.1 GWh |
|---|---|
| In 1999: | 187.0 GWh |
| Increase 1993–99: | 171.9 GWh, 1 138 % |

Ireland has few indigenous energy resources and therefore relies heavily on imported fuels. Wind exploitation offers a good opportunity for reducing fuel imports and making a significant contribution towards the country's electricity supply. Indeed, most of future renewable energy development in Ireland is expected to come from wind energy. However, some Irish developers are encountering a number of problems in obtaining planning permission for new wind farm developments. The Irish Renewable Energy Information Office provides an advice service to ensure that developers and planners receive adequate information and support to reach balanced decisions on renewable energy projects.

Kilronan wind farm is one example of a wind energy development in Ireland. It was developed by the Irish company Kilronan Wind Farm Limited. The excellent wind resource in the region (County Roscommon), together with existing access roads and power lines to the nearby local coal mines, made Kilronan the ideal location for a wind farm. It was built in 1997 and comprises 10 turbines, each of 0.5 MW. In 1998, the wind farm generated 14 GWh of electricity.

**Success factors:**

- *Political: National Sustainable Development Strategy provides a framework for renewable energy development*

Irish energy policy focuses on increasing its use of indigenous energy sources. Renewable energy can contribute towards this policy, and can also provide opportunities for rural development. In 1995 the government set a target of increasing electricity generated from renewable sources by 55 MW, from 235 MW to 290 MW. The target was achieved and extended through the 1999 White Paper on sustainable energy. This increased the target for electricity generation from renewable energy to 500 MW in the period 2000–05.

- *Legislative: The Alternative Energy Requirement (AER) provides support for renewable energy generation, through a competitive tendering mechanism*

The main support mechanism for renewable energy is the Alternative Energy Requirement (AER). This is comparable to the NFFO system in operation in the UK. It is a competitive bidding system for different renewable energy technologies. Successful developers obtain a 15-year guaranteed purchase of their power, at the project's bid price. Four AER bidding rounds have been held since 1994. Since this date, almost all new wind capacity has been achieved through AER contracts. By 1999, 10 wind farms with AER contracts were connected to the grid, totalling 63 MW, with up to 137 MW further new capacity planned.

- *Fiscal: Some fiscal support to renewables*

Limited fiscal support is available through tax relief for corporations making investments in renewable energy projects.

- **Financial: Capital grants are available for projects and for regional evaluations of wind energy resources**

Winning bidders to the AER can apply for a capital grant, supported through EU Structural Funds. A number of projects have also received support from the EU Thermie programme for demonstration projects.

Through the Business Expansion Scheme companies can access cheaper finance than is typically available from commercial banks.

- *Technological development*

There is little indigenous manufacturing of renewable energy equipment, and the opportunities for developing such an industry are limited to newer long-term technologies such as wave power, and in-site assessment activities. Investigations of Irish wind resources are actively supported, such as through the development of the Irish wind atlas by University College Dublin, RISOE (a Danish/United Nations Environmental Programme research institute) and others. Financial support has also been provided towards implementation of renewable energy feasibility studies.

- *Information, education and training: Government support for information on renewable energy*

The Renewable Energy Information Office (REIO) of the Irish Energy Centre was set up in 1995 to promote the use of renewable energy resources and to provide independent advice and information on financial, social, environmental and technical issues relating to renewable energy development. It also provides advice to the public on all aspects of renewable energy. REIO has played a leading role in identifying and addressing issues such as financing and planning, which are crucial to the successful development and deployment of renewable energy technologies. The Office is also charged with the task of disseminating the results of successful applications of renewable energy technologies. By providing ready, centralised access to such information, REIO helps reduce the cost to developers of assessing new technologies and deployment strategies. This in turn stimulates replication of successful projects which then increases market confidence. In particular the government aims to encourage smaller-scale developments of renewables in projects that can achieve local benefits. Kilronan, for example, is a local scheme helping to revitalise an old coal-mining area.

## Italy — Wind energy

*Italy's exploitation of its wind resource has been slow to take off, but its level of penetration is now rapidly expanding, especially in southern regions.*

| | |
|---|---|
| In 1993: | 4.4 GWh |
| In 1999: | 403.0 GWh |
| Increase 1993–99: | 398.6 GWh, 9 059 % |

Because of its location, in the closed Mediterranean, Italy's wind resources are not as great as those in some other parts of Europe. Nevertheless, there is still potential, especially in the Apennines Mountains above about 1 000 m, and in other locations in southern Italy.

Since the latter half of the 1990s, the Italian Vento Power Corporation (IVPC) has been active in developing a series of wind energy projects in southern Italy. IVPC linked with foreign trading partners, already involved in other wind projects around Europe, to obtain financial support for its venture. The company has now constructed wind farms at five sites, with further expansion expected. Total installed capacity is now almost 400 MW.

One of the most difficult and expensive aspects of establishing wind energy schemes in Italy can be arranging grid connections. In some regions, the existing grid infrastructure is poor (as is the case in southern Italy), and it is expensive to upgrade the grid sufficiently for it to be able to accept power from the new wind farm. In such cases, developers like IVPC often have to contribute substantially towards the connection costs of the grid operator.

In addition, obtaining permission to construct can be very lengthy — up to two years — because of the large number of permits that may be needed and a lack of clarity over the conditions that the project needs to meet.

**Success factors:**

- *Political: Enhanced political support for renewable energy is boosting opportunities for wind energy developments*

Italian energy policy aims to reduce energy import dependency and to reduce greenhouse gas emissions. Italy has implemented a series of national energy plans since the 1980s. The National Energy Plan (1988) focused on objectives and targets to 2000, including implementation of policies on energy savings, renewable energy and environmental protection. However, only limited uptake of renewable energy was achieved under these policies. New legislation in 1998–99 included aims for developing renewable energy sources to 2010 (from 11.7 mtoe in 1997 to 20.3 mtoe in 2010), based around a White Paper on renewable energy sources. The accompanying national energy programme requires regions to make regional energy plans, including policies to increase energy efficiency and to develop renewable energy, such as wind power. Nationally, new aims for wind energy uptake are identified as 700 MW by 2002, doubling to 1 400 MW by 2006.

- *Legislative: Premium-set tariffs combined with an obligation to purchase provide a stable, commercially favourable market for renewable electricity producers*

A tariff system was established in 1992. The grid operator is obliged to purchase all electricity from renewable sources, at an agreed premium rate for the first eight years of electricity production, and at guaranteed minimum rates for the remainder of the project's life.

The IVPC wind farms typically have a 15-year contract with the grid operator, which provides a premium payment for electricity output during the first eight years at ITL 202.4/kWh (EUR 0.105/kWh), after which the rate drops by about half.

Early tariff structures were complex and provided different rates for different energy sources and different times of day. Since 1992 these tariff structures have been progressively revised and simplified. In general, feed-in tariffs consist of a supplement towards the avoided costs and the higher investment costs of renewable generation compared with power from conventional sources, and apply for the first eight years.

- *Fiscal: Some support available to renewable energy projects*

Investment in wind energy (as well as in photovoltaics power) benefits from a 10 % discount on value added tax. Investment in poorer southern regions in Italy benefits from 10-year corporation tax breaks.

Funds for the financial support of renewable energy are now collected via a new carbon dioxide tax, approved in 1998. Over EUR 1.6 million were expected to be available in the first year.

## Netherlands — Photovoltaics

*Use of photovoltaics in the Netherlands has expanded steadily since 1993, mainly in small-scale and off-grid applications. A large number of grid-connected projects have also been established, and more are being developed to meet new and more ambitious national targets for PV use.*

| | |
|---|---|
| In 1993: | 0.7 GWh |
| In 1999: | 6.0 GWh |
| | |
| Increase 1993–99: | 5.3 GWh, 757 % |

Dutch businesses have been market leaders in the photovoltaics (PV) industry since the late 1980s and in particular have developed a thriving export industry. A small number of PV installations were developed in the Netherlands during the early 1990s. The Barendrecht project is an early example of a grid-connected PV installation. It was conceived in 1993 and became operational in 1996. It consists of 12 grid-connected, roof-integrated PV systems in newly built houses, with a project life in excess of 20 years. Both stand-alone and grid-connected systems have since been developed: by the end of 1998 there was an installed capacity of 5 MW of PV installations and manufacturing production of 100 000 m² of PV panels.

**Success factors:**

• *Political: More ambitious national targets have recently been established for renewable energy use*

The main driving force for encouraging renewable energy uptake in the Netherlands has been the 1997 White Paper on renewable energy. This set a target for penetration of renewable energy, subsueqently updated to 5 % of the country's gross inland energy consumption by 2010 and 10 % by 2020.

These targets and associated initiatives were not available at the time the Barendrecht project was initiated, and represent an ambitious increase compared with the levels of use of renewable energy at the time (only about 1 % in the early 1990s). Nevertheless, the Barendrecht and other earlier PV installations benefited from a range of measures that benefited renewable energy developments.

• *Legislative: Electricity supply companies are obliged to purchase power from independent power producers*

Energy policy in the Netherlands during the 1990s has built on the foundations laid by the 1989 Electricity Act, which obliged electricity supply companies to purchase and distribute electricity produced by independent power producers, such as renewable power producers.

Individual householders at Barendrecht can therefore sell power surplus to their domestic requirements to the utility. The utility buys at around NLG 0.23/kWh (EUR 0.13/kWh), which is equivalent to the price the owner would pay for conventional electricity provided from the utility. Prices are negotiated per project, however, and recent developments have now led to an average price of NLG 0.16/kWh (EUR 0.07/kWh) across the country.

Some distribution companies also accept net metering: a household's meter is allowed to run backwards as compensation for feeding the surplus electricity into the grid.

• *Fiscal: Tax structure is favourable towards renewable energy*

A regulatory energy tax (also applicable to electricity) has been in operation since 1995. This is levied on households and small and medium-sized industry; in 1999 the rate was about EUR 0.026/kWh (electricity) and EUR 0.019/m³ (gas). Renewable energy is exempted from the tax.

Companies and firms which invest in energy-saving projects (including renewable energy) can be exempted from taxation, and can benefit from claiming accelerated depreciation of investment in energy conservation equipment including PV systems. Individuals benefit from income tax exemption for investments in 'green' funds.

- *Financial: Subsidies are available for renewable energy schemes, both from public and private sources (including utilities)*

Subsidies are available both from NOVEM (the Netherlands' energy and environment agency) and from local authorities to support energy-efficient technologies, including renewable energies. In addition, utilities are obliged to invest in energy conservation or renewable energy schemes, with the result that many utilities own and operate renewable energy projects such as PV installations.

At Barendrecht, the local municipality and a private property developer initiated the project. The owners of the scheme are the homeowners themselves, although financing was provided in partnership with other local and national players, and from the utility. The owner of the house is obliged to keep the PV system for 10 years, and a contract is signed between the owner, the municipality and the utility. The system is guaranteed for 10 years after the completion date.

The Netherlands also has a number of domestic banks (e.g. Triodos) which are supportive of environmental and community initiatives, including renewable energy schemes, and are willing to invest small amounts of finance towards these kinds of projects at favourable rates.

- *Administration: Municipal authorities are responsible for implementing spatial planning, including location of renewable energy systems*

The planning process addresses the removal of any barriers to maximising solar energy production in residential buildings.

- *Technological development: Strong market development programme under way*

During the 1990s NOVEM had a strong research and demonstration support programme, assisting government research organisations and industry to develop an indigenous PV industry. Fundamental and applied research is carried out by universities and research centres, to improve cell efficiency, to investigate new market applications (both on- and off-grid), and to develop guarantees ensuring that Dutch PV products are high quality and operate to a good performance level. There is also an indigenous PV cell manufacturing capability, which by the end of 1998 was producing more than $100\,000\text{ m}^2$ of PV panels.

## Portugal — Wind energy

*Wind energy has been slow to develop, but is starting to show promise.*

| | |
|---|---|
| 1993: | 11.0 GWh |
| 1999: | 123.0 GWh |
| Increase 1993–99: | 112.0 GWh, 1 018 % |

Enernova is a subsidiary of Electricidade de Portugal (EDP) which was established to explore the wind energy potential in Portugal, and to develop the technology in the country. It was the main driving force behind implementation of wind energy schemes during the mid-1990s.

A number of wind projects were established on Portuguese islands such as Madeira in the early part of the 1990s, but the first mainland wind scheme was built in 1996. This consisted of 17 600-kW machines, totalling 10.2 MW. A second project was built soon afterwards, with a total 10-MW electrical capacity, and further schemes have followed.

The country does not have large areas with strong and predictable wind resources, but despite this a number of suitable sites have been identified through a detailed evaluation of wind resources and the subsequent development of a wind atlas of Portugal. This, combined with the increasing availability of finance for wind developments and an improvement in tariffs paid to renewable electricity producers, has led to a surge in new projects recently. However, one of the limitations to development of wind energy in Portugal is the quality of the grid infrastructure, which can result in complex and expensive connections.

**Success factors:**

• *Political: National energy programme established in support of renewable energy development*

The Portuguese Energy Programme was established in 1994 with the aim of reducing dependence on energy imports, providing reliable energy supply at a reasonable cost, encouraging energy efficiency and increasing the use of renewable energies. Regional energy plans have followed, promoting renewable energy and energy-efficiency initiatives.

• *Legislative: Premium-set tariffs combined with an obligation to purchase provide a stable, commercially favourable market for renewable electricity producers*

Since 1988 independent producers using renewable energy sources or combined heat and power have had guaranteed access to the grid at regulated prices (Decree 189/88 on independent power production). This provided a guaranteed feed-in tariff for electricity generation from renewable energy for eight years. Despite this, few wind projects came forward until financial support and tariff levels for new renewable electricity were improved in 1995 when Decree 313/95 was introduced.

Since 1999, a new feed-in law (Decree 168/99) has been implemented that provides more favourable tariffs for renewable energy, by taking account of the environmental benefits of using indigenous, non-fossil sources compared with fossil fuels, and ensuring guaranteed purchase of power from renewable generators.

• *Financial: Generation incentives for independent power producers*

The Portuguese Energy Programme in 1994 introduced financial incentives for independent electricity producers. These differ depending on the nature of the project, its size and its geographical location. Grants were available of up to 60 % of eligible costs for demonstration projects, or up to 50 % for dissemination (commercialisation) projects. Zero-interest loans were also available for up to 40 % of project costs for commercial projects. After 1997, this loan level was raised to 60 % of project costs. The Programme was funded by the European Community (Structural Funds), the Portuguese government and private investors.

More recently the government has provided finance towards grid connection costs.

- *Information, education and training: Regional evaluations of wind energy resources are available*

A great deal of work was carried out during the early 1990s to develop a wind atlas of Portugal. The Portuguese government makes information on subsidies for renewable energy projects and how to apply for them available to the public.

### Spain — Biomass power

*Biomass is an important renewable energy resource in Spain, and its use is continuing to expand rapidly, especially for power generation projects.*

| | |
|---|---|
| 1993: | 485.0 GWh |
| 1999: | 917.0 GWh |
| Increase 1993–99: | 432.0 GWh, 89 % |

Spain possesses large amounts of readily exploitable biomass resources. The household sector and various industrial sectors (pulp and paper, timber, etc.) are currently the biggest users of biomass, for thermal purposes. However, the use of biomass in power generation (including combined heat and power) is expanding rapidly. Much of this expansion is due to a combination of factors: the existence of a guaranteed market for the electricity produced at favourable rates, the availability of capital subsidies and the high level of interest shown in developing large and medium-sized biomass heat or power plants, using a variety of biomass sources as fuels.

One of the leading developers of biomass projects is ENDESA, part of a major Spanish utility. Projects include two combined heat and power plants, each generating 16 MW of electricity, fuelled by olive oil residues.

Other biomass residues used in Spain as fuel include grape/wine residues, rice husks, wood wastes and wood residues. There is also an interest in developing energy crops such as poplars or eucalyptus to fuel biomass plants, and pilot plants are being established to demonstrate the use of these fuels for energy recovery.

**Success factors:**

- *Political: Strong support for renewable energy implementation at both national and regional level*

Spain actively supports the development of renewable energy, particularly for its environmental benefits and its contribution to security of supply. The national Energy Saving and Efficiency Plan (PAEE), 1991–2000, aimed to increase the overall use of renewables by 1.1 mtoe by the year 2000, including an increase in the contribution of non-hydro renewables in electricity generation from 0.5 % in 1990 to 1.4 % in 2000. The Plan de Fomento de las Energías Renovables (2000–10) set a new target of a 12 % share for renewables in gross inland energy consumption by 2010. Biomass, and increasingly power from biomass, will be key in meeting this target.

Each of the Spanish autonomous regions has a regional energy plan, focusing on developing environmentally and economically sustainable energy provision, and containing objectives and targets for the promotion and implementation of renewable energy sources.

- *Legislative: Premium-set tariffs combined with an obligation to purchase provide a stable, commercially favourable market for renewable electricity producers*

The main driving force for support to renewable energy comes from a series of royal decrees during the 1990s on support for electricity generation from renewable energy sources, wastes and combined heat and power. The decrees guarantee the purchase of electricity from renewable sources at a premium fixed price, at 80–90 % of the average electricity tariff from conventional power sources. From 1999, electricity producers (including biomass) can receive either the fixed tariff of up to ESP 10.24/kWh (EUR 0.06/kWh) or can receive the average hourly market price of electricity plus a bonus of up to ESP 4.61/kWh (EUR 0.03/kWh). The legislation also provides for guaranteed access to the electricity grid, with agreed rates for connection.

- *Financial: State and regional subsidies available*

The PAEE provided limited (20 %) subsidies in the form of capital grants. However, the uptake of biomass projects was slower than anticipated and so capital subsidies for biomass projects were strengthened from 1996 to reach up to 30 % of eligible costs, with a further

10 % for projects developed by small and medium-sized enterprises. Each autonomous region can provide separate additional support for investment and project financing.

- *Administration: Local involvement in renewable energy planning*

Responsibility for renewable energy sources belongs chiefly to the autonomous communities (the regions). This allows each region to have authority over the various administrative procedures and over planning provisions to implement renewable energy projects. These responsibilities are closely linked with environmental obligations, and in particular with the requirements to provide environmental impact assessments for new projects.

Successful implementation of biomass projects is mostly met where collaboration at all levels of administration (local, regional and national) is achieved.

## Sweden — Biomass power

*Sweden has a long history of producing energy from its forestry resources, and biomass-based electricity, including combined heat and power, is increasing steadily.*

| Penetration: | |
|---|---|
| In 1993: | 2 113.0 GWh |
| In 1999: | 3 011.0 GWh |
| Increase 1993–99: | 898.0 GWh, 42 % |

Sweden is a world leader in the production and conversion to power of solid biomass. Most of the biomass power comes from combined heat and power (CHP) plants. In Sweden there are vast areas of forest, the fiscal system favours renewable energies and a great deal of financial and research support has been provided to biomass for a number of years. All these greatly helped biomass energy to develop as an important source of fuel for power plants. Biomass as a fuel source for CHP increased steadily during the 1990s, and particularly in recent years, and biomass power now meets about 2.5 % of Sweden's electricity needs.

**Success factors:**

- *Political: Support for renewable energy use, especially biomass*

The overall objective of Sweden's energy policy is to secure the long- and short-term energy supply on economically competitive terms, with an emphasis on sustainable development. Sweden has a policy to prevent an increase in $CO_2$ emissions, and it has also made commitments to phase out its nuclear generation capacity.

Long-term support for research and development into new and renewable energy technologies, and a greater use of renewable energy are the two principal means of achieving these aims. Biomass especially plays a vital role. Sweden has a policy objective to replace electric domestic heating with CHP or district heating systems, especially making use of biomass for fuel.

- *Legislative: Electricity supply companies are obliged to purchase power from small-scale power producers*

The liberalisation of the Swedish electricity market provides straightforward access for small independent generators to be connected to the grid. Swedish utilities were obliged to purchase electricity generated from small generators, at agreed prices. Since the last quarter of 1998, biomass power has been sold at the market price plus a temporary support of SEK 0.09/kWh (EUR 0.009/kWh) provided by the state. Small generators can also obtain discounts from grid-use costs.

- *Financial: Subsidies available to renewable energy schemes*

Investment grants are available for biomass-fired combined heat and power plants up to 25 % of total investment, which translates to a maximum of SEK 3 000/kWh (EUR 330/kWh) of electricity capacity installed. Since 1998 there has also been a technology procurement programme for renewable energy production, and biomass power projects can benefit from it. Since the last quarter of 1998, biomass power has also benefited from a temporary price support as noted in the legislative section.

- *Fiscal: Energy tax systems benefit biomass use*

Biomass is exempted from the energy tax, the carbon dioxide tax and the sulphur oxides tax. The carbon dioxide and energy taxes have helped to change the economics of new power generation and have made coal-fired CHP plants more expensive than any other option. A number of public coal-fired CHP plants have changed to fire biomass due to the introduction of the carbon dioxide and energy taxes.

Small generators are exempt from a nitrous oxide levy, which applies to generators over 25 GWh/year. The exemption applies to small generation from all fuels, not only biomass.

- *Technological development: Active development and promotion of biomass technologies*

Swedish research and development actively supports technological developments in renewable energy. Biomass research, development and demonstration receive total funding of about SEK 400 million (EUR 36 million) per year from the government. Electricity companies and other industries also provide funds. The main areas of support are combustion and conversion technologies, demonstration of pre-competitive technologies, fuel production, harvesting supply programmes and ashes recycling.

- *Information, education and training: Long history of use of biomass as fuel, benefits to key local economic actors from biomass projects*

Biomass use is well established and accepted in Sweden. Farmers and forest companies are supportive of new biomass projects because of the additional income the project will generate for them. Wood users such as sawmills also benefit because they have an additional market for their wood wastes. These actors, in particular the farmers' cooperatives, have helped to gain increasing public acceptance of biomass projects. Most important, there is a high level of environmental awareness in Sweden, particularly in renewable energies as an alternative to other energy sources, and this has often been the main force behind developing renewable energy schemes, such as biomass.

## Sweden — Wind energy

*Sweden's use of renewable energy has focused on its hydro-power and biomass resources, but it has now started to expand its use of wind.*

| | |
|---|---|
| In 1993: | 51.7 GWh |
| In 1999: | 371.0 GWh |
| Increase 1993–99: | 319.3 GWh, 618 % |

The Swedish potential for wind power is large. However, some of this is in coastal areas where wind energy development competes with other interests over land use. The size of installations has expanded considerably, and by 1999 there were 486 turbines with a total installed capacity of 220 MW.

Cooperative wind power development in Sweden has been successful. One example is an early wind power scheme (Holmbod) developed by the Vindkompaniet Swedish wind power cooperative in Gotland. This scheme comprises one 500-kW turbine connected to the grid. By the end of 1998, Vindkompaniet had installed over 80 turbines, in a variety of wind farms. Due to the success of companies such as Vindkompaniet, more than 10 % of the domestic electrical consumption in the region of Gotland is now provided by wind power.

Many of these early wind power schemes have a high level of community involvement, often being cooperatively owned by citizens living nearby. However, as the market for cooperative schemes is reaching saturation point in the windiest districts of the country, there may be fewer opportunities for further replication of the cooperative scheme involving neighbouring inhabitants. The developer is now receiving more interest from farmers and utilities than from communities, but there is also interest in cooperatives owned by citizens living across the country and not in the proximity of the project.

**Success factors:**

- *Political: Support for renewable energy*

Swedish energy policy aims to secure the country's long and short-term energy supply on economically competitive terms, with an emphasis on sustainable development. In particular, the policy during the 1990s concentrated on restructuring the energy system. There were two main areas of focus. The first was to provide support for long-term research and development into new and renewable energy technologies. The second area consisted of shorter-term initiatives to address the replacement of electricity from nuclear energy when nuclear plants are closed, through support for renewable energy, energy efficiency and district heating.

- *Legislative: Electricity supply companies are obliged to purchase power from independent power producers*

The liberalisation of the Swedish electricity market provides straightforward access for small independent generators to be connected to the grid. All Swedish regional distribution utilities were obliged to purchase electricity from small generators, at agreed prices. Small generators such as the Holmbod scheme in Gotland can obtain discounts or exemptions from grid-use costs, although they have to pay an agreed one-off connection charge and annual grid connection fees. Since the last quarter of 1998, wind power has been sold at the market price plus a temporary support of SEK 0.09/kWh (EUR 0.009/kWh) provided by the state.

- *Fiscal: Tax structure is favourable towards renewable energy*

Electricity production from small-scale renewable energy projects is favoured by lower or non-energy taxation (Act 1994:1776). For smaller and cooperative schemes, Swedish citizens are eligible for an income tax allowance of up to 15 % on new investments in renewable energy schemes. Furthermore, for cooperative operations, no income tax is payable on the share dividends up to the cost of the shareholder's normal electricity costs. This encourages investment in renewable energy schemes.

- *Financial: Subsidies available to renewable energy schemes*

Investment grants were available for wind power schemes up to 35 % for wind turbines with a capacity bigger than 60 kW. (Holmbod wind farm received a grant of 35 % of the project's development costs.) Since July 1997 investment grants have been revised downwards: up to 15 % for wind turbines bigger than 200 kW.

Wind energy schemes benefit from an 'environmental bonus' equal to the excise tax on electricity. The environmental bonus has been on average a little over EUR 0.01/kWh of electricity generated. Since the last quarter of 1998, wind power schemes have also benefited from a temporary price support, as noted in the legislative section.

- *Administration: Planning support for new wind energy developments*

Some regional planning authorities designate areas as 'suitable for wind energy production'. For Holmbod, planning was proposed and implemented at the local level by the local council. The regional planning authority instigated a detailed study of wind energy in the region. This study was used by the developer to identify a suitable site in the area. No problems were experienced in obtaining planning permission for the installation of the Holmbod turbine in Gotland.

- *Information, education and training: Active local interest in wind energy developments*

There is a high level of consciousness about environmental issues in Sweden and this has often been the main motive for developing renewable energy schemes. The entrepreneurial spirit found among many of the population and Sweden's cooperative tradition have helped to create a favourable environment for developing renewable energy. By 1998, Swedish citizens had invested about EUR 20 million in wind. Many of these investments are in cooperatives: there are about 50 already. Farmers are becoming increasingly aware of the financial opportunities, through land rentals or electricity sales, of investing in wind energy.

Vindkompaniet involved the local population in discussions at an early stage of the Holmbod project development.

# Annex 2: Study contributors

| | |
|---|---|
| Aarniala, Marjatta | TEKES |
| Åfeldt, Sten | STEM (Sweden) |
| Baguenier, Henri | European Small Hydro Association |
| Beirao, Diogo | CCE (Portugal) |
| Bertarelli , Lara | IT Power (biofuels study) |
| Bosseboeuf, Didier | ADEME (France) |
| Broom, Lis | ECOTEC Research and Consulting Ltd. |
| Cameron, Murray | European PV Association |
| Chalmers, Ewan | Irish Energy Centre |
| Da Costa, Ferreira | PROET (Portugal) |
| Daugaard, Nils | ECD (Denmark) |
| Dubuisson, Xavier | Institut Wallon (Belgium) |
| Egger, Christianne | ESV (Austria) |
| Fock, Martin | Centre for Biomass Technology (Denmark) |
| Fraga, Juan | EUFORES |
| Garzon, Dina | SODEAN (Spain) |
| Grulois, Christophe | ERBE (Belgium) |
| Gutiérrez, Cristina | European Forum for Renewable Energy Sources (Spain) |
| Hackstock, Roger | EVA (Austria) |
| Hämmerle, Kurt | Energieinstitut Vorarlberg |
| Hemmelskamp, Jens | European Commission — Joint Research Centre — Institute of Prospective Technological Studies |
| Jakelius, Stephan | STEM (Sweden) |
| Kellett, Paul | Irish Energy Centre |
| Kuebler, Knut | European Commission, Energy and Transport Directorate General |
| Lemming, J | Danish Energy Agency (Denmark) |
| Lorenzoni, Arturo | Bocconi University (Italy) |
| Mandrup, Klaus | Danish Energy Agency (Denmark) |
| Nurmi, Markku | Ministry of Environment (Finland) |
| Oliveira, Pedro | AMES — Agência Municipal de Energia de Sintra (Portugal) |
| O'Regan, Fionna | Irish Energy Centre |
| Rakos, Christian | EVA (Austria) |
| Roubanis, Nikos | Eurostat |
| Sanchez, Manuel | European Commission, Research Directorate General |
| Simader, Gunter | EVA (Austria) |
| Swens, Job | NOVEM (Netherlands) |
| Thorson, Ola | STEM (Sweden) |
| Ullerich, Dirk | European Commission, Energy and Transport Directorate General |
| Vainio, Matti | European Commission, Environment Directorate General |
| Volpi, Guilio | World Wide Fund, European Policy Office |
| Wagner, Andreas | German Wind Energy Association |
| Yordi, Beatriz | European Commission, Energy and Transport Directorate General |
| Zervos, Arthuros | European Renewable Energy Council |

# Annex 3: Study reviewers

| | |
|---|---|
| Åfeldt, Sten | Swedish National Energy Administration (Sweden) |
| Bosch, Peter | European Environment Agency |
| Carritt, Tony | European Environment Agency |
| Egger, Christianne | Energy Agency, Upper Austria (Austria) |
| Fontana, Michele | European Environment Agency |
| Froste, Hedvig | Swedish Environment Protection Agency (Sweden) |
| Garzon, Dina | Sodean (Spain) |
| Geissler, Michael | Berlin Energy Agency (Germany) |
| Horrocks, Peter | European Commission, Environment Directorate General |
| Huntington, Jeff | European Environment Agency |
| Jol, Andre | Air Emissions Inventories Project Manager, European Environment Agency |
| Larsen, Jens H. | Copenhagen Environment and Energy Office (Denmark) |
| Person, Susann | Swedish National Energy Administration (Sweden) |
| Rosenstock, Mayfred | European Commission, Environment Directorate General |
| Sanchez, Manuel | European Commission, Directorate General Research |
| Säynätkari, Tapani | Finnish Environment Institute (Finland) |
| Singer, Stephan | Climate and Energy Unit Head, World Wide Fund |
| Smith, Ian | European Environment Agency |
| Stanners, David | European Environment Agency |
| Vos, Hans | European Environment Agency |
| Zapfel, Peter | European Commission, Environment Directorate General |
| Zervos, Arthuros | European Renewable Energy Council |

# Annex 4: Offshore wind

The market penetration of offshore wind energy has not been evaluated in this study because the technology is only at an early stage of implementation and there are few truly offshore plants in operation. Nevertheless, it is estimated that there is an exploitable potential for offshore wind of more than 3 000 TWh/year (Hassan and Lloyd, 1995) and there is an increasing level of activity in many Member States to exploit this resource.

Belgium, Denmark, Germany, Ireland, the Netherlands, Spain, Sweden and the UK are all at different stages in the exploitation of their offshore wind resources. Blyth wind farm in the UK and Middlegrunden near Copenhagen in Denmark were constructed during 2000.

Are there lessons to be learnt from previous experiences of renewable energy developments onshore — and in particular from the case studies evaluated in the present study — that could be relevant to the offshore situation?

**Political**: This emerging technology is still more expensive than most established land-based renewable energy technologies, and requires some form of financial subsidy to bridge the stage between early demonstration turbines to full-scale commercialisation. Offshore wind farms are considerably larger (wind farms of up to 100 MW are envisaged) than most of their onshore equivalents. The level of commitment from national governments to support offshore wind energy development would therefore need to be strong and over a long period, in order to provide a stable and encouraging environment for long-term investment from the private sector.

**Legislative**: By and large, the success of onshore wind generation has been driven by the long-term stability and economic viability provided to the onshore generators by output support mechanisms, in particular feed-in tariffs. However, the move towards an EU-wide liberalisation of electricity markets makes the establishment of new feed-in or competitive tendering arrangements for offshore wind more complex than during the 1990s. Competitive electricity markets, and cross-border trading in electricity, are not compatible with national feed-in arrangements in the long term. New power purchase guarantees and output support mechanisms may therefore have to be established to ensure long-term stability and economic viability for offshore wind. This could be a useful guiding principle in the establishment of any new arrangements.

**Financial**: The extensive experience gained from the development of onshore wind developments has increased the level of confidence of developers as they expand into the new offshore market. This confidence is also seen in banking and financial institutions, which are willing to invest in this emerging technology without significant levels of capital subsidy as long as there are appropriate revenue forecasts available (through power purchase arrangements).

**Fiscal**: As a non-fossil source of energy, electricity from offshore wind is eligible for exemptions or rebates from energy or carbon taxes. As more and more Member States implement environmental taxes, this will continue to improve the competitiveness of renewable energy in comparison with fossil energy sources.

**Administrative**: Offshore wind installations open up new aspects of planning permits. Offshore turbines interact with conservation, fishing, tourism and shipping interests. The siting of the turbines would therefore have to be carefully considered. Active support from local administrations will be important in this respect.

**Technological development:** Offshore wind turbines draw on technological capabilities from both the onshore wind industry and the offshore installations industry (particularly from the oil and gas industry). Much research and technological development is being carried out by turbine developers, especially through a step-by-step approach to installations, with considerable monitoring and evaluation taking place at all stages. Government financial support often focuses on feasibility studies and other non-technical support. This partnership approach between government and industry helps to ensure that the systems and approaches being taken to develop this new technology are robust and will ultimately be technically and financially successful.

**Information, education and training:** Public support for offshore wind developments is critical if this technology is to be widely accepted. Most offshore developers have carried out extensive public consultations and information campaigns to raise awareness and obtain public support. The Copenhagen wind farm, for example, ensured public support not only through consultation but also by enabling local people to purchase shares in the project through a cooperative.

Offshore wind is projected to make an important contribution towards EU policy and targets for renewable energy deployment in the next 5–10 years. Its progress is benefiting from the experiences gained in overcoming barriers and in identifying potential success factors for deploying renewable energy technologies (especially wind energy) on shore during the 1990s.

# Glossary

| | |
|---|---|
| CHP | Combined heat and power |
| $CO_2$ | Carbon dioxide |
| Eurostat | Statistical Office of the European Communities |
| GW | Gigawatt — a unit of electrical power, equal to 1 000 000 000 Watts |
| GWh | Gigawatt hour — 1 GW operating for 1 hour |
| ktoe | Thousand tonnes of oil equivalent |
| kWh | Kilowatt hours |
| kWp | Kilowatt peak |
| $m^3$ | Cubic metre |
| mtoe | Million tonnes of oil equivalent |
| MW | Megawatt — a unit of electrical power, equal to 1 000 000 Watts |
| MWp | MW at peak power (photovoltaics functioning in optimal sunlight) |
| $NO_x$ | Nitrous oxides |
| PV | Photovoltaics |
| R&D | Research and development |
| RD&D | Research, development and demonstration |
| $SO_2$ | Sulphur dioxide |
| VAT | Value added tax |
| Wp | Watts at peak power (photovoltaics functioning in optimal sunlight) |

| | |
|---|---|
| Final energy consumption | Energy consumption by final user — i.e. which is not being used for transformation into other forms of energy. |
| Gross inland energy consumption | A measure of the energy inputs to the economy, calculated by adding total domestic energy production plus energy imports minus energy exports, plus net withdrawals from existing stocks. |
| Gross electricity consumption | Domestic electricity production, plus imports, minus exports |
| Electricity generation | Electricity production, i.e. total amount of electricity produced. Note that at present there are only minimal amounts of electricity from renewable sources which are imported or exported. As a result, the terms gross electricity consumption and electricity generation can be considered to be equivalent for electricity from renewable sources. |